Othermindedness

Studies in Literature and Science
published in association with the
Society for Literature and Science

Othermindedness
The Emergence of
Network Culture

Michael Joyce

Ann Arbor

THE UNIVERSITY OF MICHIGAN PRESS

First paperback edition 2001
Copyright © by the University of Michigan 2000
All rights reserved
Published in the United States of America by
The University of Michigan Press
Manufactured in the United States of America
⊗ Printed on acid-free paper

2004   2003   2002   2001   4   3   2

A CIP catalog record for this book is available from the British Library.

Library of Congress Cataloging-in-Publication Data

Joyce, Michael, 1945–
     Othermindedness : the emergence of network culture / Michael
Joyce.
        p.   cm. — (Studies in literature and science)
     Includes bibliographical references (p.   ).
     ISBN 0-472-11082-9 (alk. paper)
     1. Fiction—Authorship—Data processing.   2. Experimental
fiction—History and criticism—Theory, etc.   3. Authors and
readers—Data processing.   4. Fiction—Technique—Data processing.
5. Creative writing—Data processing.   6. Storytelling—Data
processing.   7. Literature and technology.   8. Hypertext systems. I.
Title.   II. Series.
PN3377.C57J69   1999
808.3'0285—dc21   99-6770
                                                                      CIP

ISBN 0-472-08843-2 (pbk. : alk. paper)

O Caro, for you

The *arrivance* of love is the genesis of the interior.
—Cixous, "What is it o'clock?"

Acknowledgments

As always all thanks begin with family, my sons Eamon and Jeremiah foremost, both of whom having come now of an age where they read at least some of my work with patience and even, at times, surprising and consoling interest. My brothers and sisters by and large continue to think me the big brother whose mysterious doings must somehow make sense. Carolyn suffuses.

A dozen colleagues, near and far, save me from stupidity and gently nudge me toward whatever comes next despite my reluctance and fundamental reticence. I think of Jay Bolter as an unflappable and endlessly patient teacher, Howard S. Becker as my *sensei*, and Kate Hayles as among the muses. I have been lucky in invitations elsewhere, both virtual and actual, which have brought me into the presence (and under the goading scrutiny) of young theorists and artists like Ana Boa-Ventura, Carolyn Guertin, Hilmar Schmundt, and others whose push and flash has supplemented but not supplanted the sustaining force of the now decade-old TINAC group, J. Yellowlees Douglas, Nancy Kaplan, John McDaid, and Stuart Moulthrop. At the University of Michigan Press, LeAnn Fields continues to be a generous and encouraging editor, and the production team makes this task a pleasure. At Vassar, Barbara Page and Paul Kane especially have sustained, supported, and challenged me. My student assistants, Lorrie Rivers and, most of all, Chris Guenther have worried and worked over this manuscript with me. Matt Hanlon has gone from being the English intern to a gifted web designer and collaborator out in the actual wide world. The small band of student consultants in my working group in the Center for Electronic Learning and Teaching, including Ken Bolton, Cristina Carp, Roland Gauthier, Vance Lin, Andrew Rosenberg, Lucas Smith, and Sarah Torre have instanced the emergence of network culture, as have my extraordinary and fearless students. To the latter go my most fervent thanks and my deepest acknowledgements for all we have learned together.

Finally the list of permissions that follows does not do justice to the permission to risk, to speculate, and to tentatively formulate that various editors and journals have offered me.

Grateful acknowledgment is made to the following editors, publishers, and journals for permission to reprint previously published materials.

"(Re)Placing the Author: 'A Book in the Ruins.'" In *The Future of the Book,* edited by Geoffrey Nunberg and Patricia Violi. Copyright © 1996 Geoffrey Nunberg and Patricia Violi. Reprinted by permission of the University of California Press. This chapter was originally given as a talk on July 28, 1994, in San Marino at "The Future of the Book," a conference cosponsored by the International Center for Semiotic and Cognitive Studies and the Rank Xerox Research Centre of Grenoble.

"MOO or Mistakenness." *Works and Days* 25–26 (summer–fall 1995). Reprinted by permission of David Downing, editor.

"New Stories for New Readers." In *Page to Screen: Taking Literacy into the Electronic Era,* edited by Ilana Snyder. Melbourne: Allen and Unwin Pty. Ltd., 1997. Reprinted with permission of publishers. Parts of this essay were originally given under the title "Writing in the Middle Voice: Vernacular Coherence, or, Seeing Ourselves in Where We Are," plenary talk for the Mid-Atlantic Alliance for Computers and Writing Conference, February 9, 1996, George Mason University, Virginia.

"The Lingering Errantness of Place, or, Library as Library." In *The Emerging CyberCulture: Literacy Paradigm and Paradox,* edited by Stephanie Gibson and Lance Strate. Hampton Press, 1999. Reprinted with permission of publishers. This essay was originally given in slightly different form as an invited talk for the ACRL-LITA Joint Presidents Program, American Library Association, 114th Annual Conference, Chicago, June 26, 1995.

"Beyond Next before You Once Again: Repossessing and Renewing Electronic Culture" first appeared in *Passions, Pedagogies, and 21st Century Technologies: Confronting the Issues,* ed. Gail Hawisher and Cindy Selfe (Logan: Utah State University Press, 1999).

"Songs of Thy Selves: Persistence, Momentariness, Recurrence, and the MOO." In *High Wired: On the Design, Use, and Theory of Educational MOOs*, ed. Cynthia Haynes and Jan Rune Holmevik. Ann Arbor: University of Michigan Press, 1998.

"One Story: Present Tense Spaces of the Heart." In *In Memoriam to Postmodernism: Essays on the Avant-Pop*, edited by Mark Amerika and Lance Olsen. San Diego: San Diego State University Press, 1996. Reprinted with permission of publishers.

"Nonce upon Some Times: Rereading Hypertext Fiction." *Modern Fiction Studies* 43, no. 3 (1997): 579–97. Copyright © 1997. Purdue Research Foundation. Reprinted by permission of Johns Hopkins University Press. Collected in *Second Thoughts: A Focus on Rereading*, ed. David Galef. Detroit: Wayne State University Press, 1998.

"On Boundfulness: The Space of Hypertext Bodies." In *Virtual Geographies: Bodies, Space and Relations*, edited by John May, Phil Crang, and Michael Crang. London and New York: Routledge, 1998. Reprinted with permission of the publishers.

"Forms of Future" was originally given in slightly different form as a keynote talk at "Softmoderne3, Festival of Network Literature," Berlin, September 1997; and in still another form at "Transformations of the Book," Massachusetts Institute of Technology, October 1997.

Contents

Othermindedness: The Emergence of Network Culture

There are good reasons—besides euphony or the foolish consistency that leads one to name successive hypertext fictions *afternoon, a story* and *Twilight, a Symphony*—to follow one collection of essays with mind in its title, *Of Two Minds,* with another mindedness here.

If the earlier book summoned and reified a commonplace saying in order to highlight in both computer pedagogy and poetics the sense of perspectivilization—the oscillation between looking at and looking through that Jay Bolter and Richard Lanham have identified as characteristic of the computer, then the current book means to suggest a latter development in the emergence of what I have called network culture. Network culture is not, to my mind at least, the same as network*ed* culture, which is quite a different thing, less projective, more conventionally social. Network culture is an othermindedness, a murky sense of a newly evolving consciousness and cognition alike, lingering like a fog on the lowlands after the sweep of light has cleared the higher prospects. The same or a like fog increasingly seems to cling in the folds of the brain. We ache with it, almost as if we could feel the evolution of consciousness in the same way a sleeping adolescent feels the bone ache of growing pains as if in a dream.

My focus in this collection, as I say in a chapter that follows, is to summon an othermindedness that is less a focus on the other than upon our mindedness. Network culture, if there is to be such a thing, calls us to a new mind, one in which we must not merely affirm seemingly passive choices but find a ground upon which to do so; where we must not only insist upon a natural warrant for virtual worlds but also continually articulate their differences from and affinities with the world we inhabit in nature; where we must not only identify the continuity of embodied spirit that we increasingly represent as alternate selves, but affirm it.

To say that network culture involves a newly evolving consciousness and cognition is not to make a claim for technological determinism. We make the world new as we see it, but we increasingly see it in devices that seem (or seek) to contain us. Nor is it to give over to the closed-circuit determinism of the evolution of machine consciousness in devices that contain latter-day versions of themselves, implicit in the evolving network of the web, explicit among researchers involved in A-life (artificial life). Yet we will increasingly have to contend with claims for distributed life and consciousness even as we struggle to understand what that might mean for our own prospect and continuity.

I had dinner in recent months with a physicist and a mathematician, both young, both Germans, both computer savvy, who reflected calmly over red wine and samosas about their certainty that computers would before long share consciousness with us. "If a computer can compose music a music critic believes to be a lost minor work of Mozart, then there is no magic which we can assign to human consciousness," said the physicist, otherwise a cultured and witty man. I said I thought there was. He acknowledged my belief and renewed his own, saying, "That's why you are artist and why also I am a scientist."

Yet if we inhabited, however unwittingly and unwillingly, the old two-cultures polarity, it was more the polar sense of Arctic bears circling each other in shared blindness, shuffling in the face of a frozen and bright expanse where we neither could see through nor make out its features. Increasingly any two or twenty or any number of cultures inhabit what seems one white space, its features burnished smooth by shifting light.

Thus I mean also another othermindedness in which I know less and less as this cultural shift takes place before me, as if the sweep of light not only transforms the vista but all that brought me here and that no longer seems to sit at my back. "I was so much older then, I'm younger than that now," Bob Dylan sang of all our back pages. In this collection, therefore, there is little of the migration of text from one essay to another that figured so prominently in my earlier collection, not because a point has been made—it was never a point but a practice—"increasingly characteristic of the late age of print," as I wrote then—in which "electronic texts . . . moved nomadically and iteratively from one talk to another, one draft to another, one occasion or perspective to another." That I have not done so now does not signal a lapsing of this practice but rather an increasingly fragmented sense of myself as a creature of network culture, a fragmentation that makes me less likely to think that a given text could migrate comfortably among occasions or shift perspectives.

This is not to say that I have retreated from the hypertextual, only lived into it far enough that I no longer find satisfying factors in its shifting features, no longer feel certain in my life of even the transitory closures that sustain me in my art. I console myself with knowing that no one around me seems to have any more satisfactory answers: neither the media giants who would make presumptive claims on network culture if they could decipher its nature and whose business it is to do so, nor the networked culture of artists, critics, theorists whose glimpses feed me and whose gestures I follow like a man dancing with shadows. I still throw in my lot with the latter, of course, because, as my prefatory meditation here notes, I love shadows and trust outsiders.

As a final note I should account for the use of *emergence* in the subtitle of this collection, a usage that certainly is calculated to associate these essays with the widespread interest in emergent behavior, chaos theory, and the like within the humanities. This is to say I do not deny the association between the thinking here and those moments of perturbation (explicitly discussed in the first chapter) that shift even shifting stabilities into some other state.

"Don't you always want to stay in this state?" a former colleague's husband asked postcoitally on their wedding night. "Michigan!?" she said incredulously. I mean emergence in the way she intuited the end of their marriage in its beginning. Change lies in things but is disturbed unpredictably; in the course of the disturbance not only do the things change but change itself does. Changing change, a term I have used previously, constitutes emergence. Like any buzzword, this one is meant to buzz, but not like a beeper as much as a bumblebee banging on glass. In my recent thinking I haven't felt any higher calling, cellular, satellite, or otherwise. In lieu of knowing with any certainty, I've chosen to flit for these last years, looking for color, longing for what sweetness the undifferentiated haze offers, inevitably lying blind and stunned. Despite their inevitable polemical and poetical turns, these essays are offered in that spirit. The view from here, abuzz.

That said, the collection here is fairly straightforward, eleven chapters bracketed by a meditative prelude and coda, with a similarly meditative intermezzo interspersed just beyond the midpoint. The essays here, like those of the last collection, continue to take the form of what I have called in *Of Two Minds* a theoretical narrative, both a narrative of theory and a text theoretically at least a narrative and thus not unlike what Gregory Ulmer calls "mystory." Unlike my first collection, there is not a lot of explicit talk about pedagogy, though teaching and the thinking of my students flow through here like a subterranean stream. A number of essays consider the shifting nature of the library as an instance of emergence. Otherwise there is much the same (perhaps strange) range of subjects and interests familiar to readers of my previous collection, from geography to interactive film, from MOOs and other virtual spaces to life along a river. What is perhaps new is a distrust of claims for both newness and the next as well as a recurrent insistence upon grounding our experience of the emergence of network culture in the body.

Not so much new here as newly limned by the act of collecting previously published essays are the pervasive autobiographical elements throughout these chapters. I can only hope that these elements serve as expressions of my repeated belief "that the value of our presence as human persons in real place continues as a value *not despite but because of* the ubiquity of virtual spaces. Our embodiment graces actual and virtual space alike with the occasion for value." The prelude, "Screendoor: A Meditation on the Outsider," means to be a memoir of both actual and virtual space.

Chapter 1, "(Re)Placing the Author: 'A Book in the Ruins,'" was prepared for the conference on the future of the book, sponsored by Xerox PARC (Grenoble) and Umberto Eco's Centro Internazionale di Studi Semiotici e Cognitivi in San Marino. I have sometimes fancied it a parody of close reading, reading a poem of Milosz as a meditation on electronic text. It begins with the premise that whatever the future has for some time required of us incorporates the past, or rather pulls it through as through the wormhole of a singularity. It proceeds through a reading of Milosz and Sanford Kwinter's essay on Boccioni to consider the spaces our minds create, whether in poetry or varieties of electronic experience.

Chapter 2, "MOO or Mistakenness," builds upon the notion of "the interdeterminability of points of perception" that chapter 1 offers as the fundamental impediment against any satisfying virtual reality. Among the most self-reflexive, hyper, and hypertextual essays here (and one of the few in which nomadic texts from previous chapters appear), it attempts to approximate in linear prose the experience of MOOs as textual virtual realities. It is not surprising, the chapter argues, that the MOO is a locale of mistakennesses. "There is a general feeling that the MOO is a mistake of technological history, a developmental lacuna, a place marker, an interregnum in the immanent hegemony of the postalphabetic image . . . that the MOO moment is temporary and that soon the image will either rob us of the power—or relieve us of the burden—of language."

Chapter 3, "New Stories for New Readers," the first occurrence of my notion of othermindedness, picks up from the locale of a MOO class to consider "stories of technological presence and multiplicity [and] how used to them we are." It extends this inquiry to a critical consideration of the world wide web and suggests a new voice for our interactions, one grounded in age-old values and a forgotten syntax, a middle voice where we begin to see ourselves in where we are and encourage responsibility for our choices. Who we are, the chapter suggests, "is predicated upon a necessary and creative scrutiny of the things we are used to, especially as they have to do with our understanding of differences: between the virtual and the embodied, between the lasting and the transient, between the rare delights of human community and presence and the universal promise of access and equality."

The first of the essays about the library in the electronic age, chapter 4, "The Lingering Errantness of Place," is a mediation on "error and wander" as both ways of knowing and signs of the emergence of a new mind. The library is considered as "a profession of the value of human multiplicity, proximity, and community," a locale where "in the face of . . . voracious newness . . . we might interpose the lingering errantness of place, the heterogeneous practice of culture as the experience of living in a place over time."

Chapter 5, "Beyond Next before You Once Again: Repossessing and Renewing Electronic Culture," is an explicit meditation on place in the face of an emerging electronic culture seemingly too ready to discard not only place, but body and history. It borrows as its subtitle the name of Sherman Paul's collection of "essays in the Green American Tradition," *Repossessing and Renewing,* as a conscious nod and a continued memorial to my mentor, who late in his life offered me the grace of affirming that my hypertextual experiment was for him within the Green Tradition. This essay intends a gesture toward what comes beyond next, which is nothing less than what is before us: ourselves as expressed within time and space.

The next chapter, chapter 6, "Songs of Thy Selves: Persistence, Momentariness, Recurrence, and the MOO," is likewise both an elegy and a meditation on place, looking at the MOO (and poetry alike) as "a conscious attempt at a proximate geography, a claim for the transcendence of the virtuality of language over the mortality of the body" wherein, "like any poetic text, the MOO aspires to moral discourse."

The intermezzo, "One Story: Present Tense Spaces of the Heart," suggests a similar inclination toward moral discourse in hyperfiction, where what we read is "the difference between the desire and the trace [and thus] how the forms of things mean." Originally published as part of a (premature, I think) "Memoriam to Postmodernism," this meditation situates hyperfiction within a weave of texts from the Maya, to Gertrude Stein, to contemporary feminist fiction and poetry all "trying to see a truly participative, a multiple, fiction."

Chapter 7, "Nonce upon Some Times: Rereading Hypertext

Fiction," considers the work of two hypertext fiction writers, Mary Kim Arnold and Shelley Jackson, as well as hypertext poet and theorist Jim Rosenberg as instances of how "hypertext only more consciously than other texts implicates the reader in writing at least its sequences by her choices [and how hypertext] more clearly than other texts seems to escape us before we have it formed into an understanding we might call a reading."

"On Boundfulness: The Space of Hypertext Bodies," chapter 8, originally written as a contribution to a collection of essays on geography and cyberspace, considers those spaces that are both within and somehow simultaneously outside the space of the text, most notably our bodies. Another highly self-reflexive chapter, it suggests that "the gesture of the parenthetical, the dialectic, the thematic, the rhythmic, the fugal, the isobaric, the metonymic, the list, the link, the litany . . . constitute the space of hypertextuality. Boundfulness, in this sense, is space that ever makes itself."

Both chapter 9, "Forms of Future," and chapter 10, "Paris Again or Prague: Who Will Save Lit from Com?" look to contemporary Europe as an occasion for reflections upon new media that invert those of Tocqueville, moving beyond democracy in America and, for good or ill, toward technocracy in networked Europe. Chapter 9 suggests that "the emergence of a truly electronic narrative art form awaits the pooling of a communal genius, a gathering of cultural impulses, of vernacular technologies, and most importantly of common yearnings that can find neither a better representation nor a more satisfactory confirmation than what electronic media offer." Building on a meditation of Berlin as a locale for "the constant blizzard of the next," the chapter argues that "we must nonetheless find our way through both our own private histories and the cumulative history of our cultures . . . a history of our making and our remembering alike."

Chapter 10 suggests that Prague is, if only in the smoke-wreathed icon of its poet and playwright president, Paris again, the new, perhaps the last, republic of words, a new, perhaps the last, gasp of lit before com. Looking at a contemporary German interactive video artist and a contemporary Irish writer and visual artist, the chapter argues that "not just lit but com as well depend upon our ability to interrupt the flow of nextness with a sustaining sense of the ordinary."

The last chapter is such a history of making and remembering alike. This essay, "My Father, the Father of Hypertext, and the Steno . . ." begins as a reading of Vannevar Bush's seminal 1945 essay "As We May Think." This reading is both interspersed and followed by a theoretical narrative that centers on three figures, my father, an armchair philosopher-scientist and a real-world photographer and steelworker; the father of hypertext, as Bush is often considered since his "Memex" essentially outlined the scholars' workstation and the world wide web; and the anonymous steno, sometimes called the Typist, who recurs as a figure in Bush's essay and whose "impulses . . . flow in the arm nerves . . . [and] convey to her fingers the translated information which reaches her eye or ear." Bush dreams of intercepting these impulses in a gesture of what seems both cyborgization and however unconsciously figured sexual imperialism. Perhaps, this essay will suggest, the war against memory is a struggle against embodiment and birth itself, the double portal of memory.

The coda, "Portrait of the Artist as a Search Engine Entry," also ends with a memoir of my father as well as of me as father. While it suggests that, for now at least, "the most likely portrait of you that would emerge if you got run over by a laundry truck wouldn't come from the internet but from the contents of your wallet," it nonetheless finds evidences of the emergence of network culture in the "naive mix of coherence and happenstance left out for a world to see" both apparent on the web and, one hopes at end, throughout this collection as well.

Screendoor: A Meditation on the Outsider

> On the one hand there are those who waste away in agonizing struggle between what no longer is and what will never be—the followers of neutrality, the advocates of emptiness; they are not necessarily defeatists, they often become the best of ironists. On the other hand there are those who transcend: living neither before nor now but beyond, they are bent with a passion that, although tenacious, will remain forever unsatisfied. It is a passion for another land, always a promised one, that of an occupation, a love, a child, a glory. They are believers, and they sometimes ripen into skeptics.
> —Julia Kristeva, *Strangers to Ourselves*

Every outsider is also in. Or only. Whether within the solitary husk of self or the categorical position she is put in: fat or thin, lonely, poor, alien, bright, dreamer, one of them, belongs to him, used to be, wants to be, will end up, at the window.

I spent much of my adolescence looking out the screen door. Even in winter I looked out, swinging the heavy front door back on its hinges, peering through the fog of breath I made there, the frost, the same skim of dirt year to year, never washed off when the screen came down and the outer storm on before the season. It became something of a family joke or, sometimes, an annoyance, in the way a father will be annoyed about the relentlessly habitual in an adolescent son.

Inside I was always out. Beneath the spill of light from the utility pole toward the corner where the Protestant church was, scratchy bricks and a smooth stoop of weathered gray concrete, its nave a cavern, its sad yard right field when we played baseball on the street. Sometimes, as someone walked by, the stub of a cigarette went up like a small, sad rocket launched from between forefinger and thumb toward the gutter or yard. Sometimes packs of children went by, arm in arm, tough, tender, already on their way to what they would become, futureless and featureless replica of mother or father. It was an Irish ghetto, Irish-American, though we didn't observe the hyphen or didn't admit the ancestor culture could be something on its own. We were the Irish and they were what was left (or we didn't know any better). Outside in. Arrogance is the expression of an outsider's vulnerability. That and loss. Hope is only loss seen from within.

What was I looking for? Certainly there was no image for my expectation, no white horse, headstrong and snorting, shoes clattering on the brick streets, flanks silver with sweat under the streetlight. Nor a goddess dragging a two-wheeled wire shopping cart like a chariot behind her, chewing gum, the outline of her bra straps through the back of her T-shirt, her boyfriend whistling, swearing, launching cigarette butts, or absent altogether.

Waiting. I was patient then, I think, or a certain kind of bored that passes for patience. A knowledge that there would never pass here any person or thing that could satisfy my wanting. Nor that there was any alley, exit, shortcut, gate that would take me farther than my eyes could. Cricket chirps and passing traffic, someone's shout, a faraway laugh, moths or june bugs clattering against the mesh of the metal screen.

It must have pleased me, poised there, both in and out, waiting for what would not, what could not, come. This is of course the language of sexuality; I am not unaware that an adolescent boy at a door is inevitably an image of unassigned longing. (Were there flowers in my neighborhood? There was an alley beside the church choked with Queen Anne's lace and smelling therefore of dirt and carrots, where once we smoked and once someone's older sister languidly, bored, peeled back her swimsuit top to show us a breast. I think she wanted us each to pay something to see it. But were there other flowers, sweeter, lilacs? locust? rhododendron fumes? intoxicating and nearly sickening honeysuckle nights in August? Who can say, the past is gone.)

Sometimes, I am certain, though I can recall no example, I thought things, wonderful things, brave things, eloquent, poetic, surprisingly wise even to my own eyes. Ideas. I remember nothing. There were occasional cars (they would have had fins or bulbous fenders, glasspack mufflers, turquoise paint). I wasn't an only child (not in an Irish house!), rather the oldest of eight but I cannot remember much of their noise behind me. Perhaps that is why I looked out: to be within myself, briefly outside them. Perhaps I meant to leave them an image of myself, looking outward, dreaming, bravely, meditative, melancholy, bound elsewhere.

I can remember the scent of the screen, rusty and faintly mildewed, and the night air sometimes sweet as fog, other times a humid swelter. In those days I wore copious amounts of aftershave: Canoe for awhile and then some lime fragrance. In those days there were still great elms along the street, the light pooling beneath them.

I think this is an image of how I feel about the computer screen as well. We seem lonely to me there, I have written elsewhere (outside in) about the world wide web. Sometimes when I say this it annoys others, in the way a father is annoyed by a son's denial of the obvious; sometimes others merely pity me, the way passers-by mock someone standing in the shadow of an open door, moving on, not really knowing or caring who he is. Looking out, uncertain, scanning from pooled light to deep shadow, from passing lights to solitary walkers along the sidewalk or a boisterous group, half drunk and shouting down the center of the street. It is good to be there and here at once, but lonesome nonetheless, bereft, lost, grasping. Maple seeds helicopter down into the brick street and are swept to the gutter by the breeze or the billow from passing cars.

Chapter 1

(Re)Placing the Author: "A Book in the Ruins"

> A dark building. Crossed boards, nailed up, create
> A barrier at the entrance, or a gate
> When you go in. Here, in the gutted foyer,
> The ivy snaking down the walls is wire
> Dangling. And over there the twisted metal
> Columns rising from the undergrowth of rubble
> Are tattered tree trunks. This could be the brick
> Of the library, you don't know yet, or the sick
> Grove of dry white aspen where, stalking birds,
> You met a Lithuanian dusk stirred
> From its silence only by the wails of hawks.
> ("A Book in the Ruins" Milosz 1988)

The poet stands in the ruins, it is the modernist moment. But, no, this is not what we see. The poet makes his way into the ruins, and in so moving the movement in itself reads barrier as gate. What he reads he writes. Against the unassigned space of the dark building—let us call this space the screen—crossed boards read as sign and juncture, axis and nullification, light on light. A gutted foyer no longer opens to another space but is the space of its own opening. Yet in movement the outward dark gives way to memory and metonymy, to multiplicity, the one continually replacing the other. In this crossing "here" and "there" are interwoven as if ivy. Or wire. Shifting light, dusk or dawn, yields what I have called elsewhere the momentary advantage of its own awkwardness, a slant illumination in which we are allowed to see each thing each time interstitially, in the moment before it assumes its seamlessness in the light of day, the dark of night. The electronic age now enjoys the time of awkwardness before the age itself disappears along with the mark of its name into the day to day of what at Xerox PARC they call "Ubicomp" (Rheingold 1994, 93), ubiquitous computing. For now, in the shifting light of the present awkwardness, even though the figures move asymmetrically (ivy is wire, column is tree), each thing nonetheless remains itself even in the displacement, the chiasmus, of its form. What's changed is not the thing but its placement. Print stays itself, electronic text replaces itself. Electronic text is as apt to evolve before it forms, as apt to dissolve before it finishes. On the screen it takes our constant and attentive interaction to maintain even the simulacrum of static text. The future, too, requires as much of us, and has for some time.

There is a play on words in this formulation, since whatever the future has for some time required of us incorporates the past, or rather pulls it through as through the wormhole of a singularity. The future won't stay still but instead keeps on replacing itself. The page becomes the screen, the screen replaces the page. We could call this placement history. Electronic texts present themselves in the medium of their dissolution: they are read where they are written, they are written as they are read. What "this could be," the poet tells us, "you don't know yet." The boy who stalked birds himself is stalked, circled by history. Only wailing hawks have the perspective of memory, and in memory they are said to circle a sick grove, although one whose locus, we know extratextually, is the golden age, the childhood land, of the poet who here—the year is 1941 and it is Warsaw—is a young man, a janitor in a bombed library, who "carted books from the University Library to the National Library, and from the National Library to the Krasinki Library" (Milosz 1987, 275).

I want to speak carefully. In reading this poem of Milosz as a meditation on electronic text, I do not wish to appropriate the horror of this scission, to trivialize what the space of the Warsaw library opens into or what, in slicing through, it replaces. Alexander's sword cut through the topological manifolds that gave meaning to the Gordian knot—we could call these twined strands ivy or wire, word or superstring—and in this movement he both read and wrote his own fate. "There was a sense then that all cultural values were undergoing tremendous destruction, the total end of everything," says Milosz (1987, 275). The hole replaces the whole, they carted books behind not before the storms. And yet the book in the ruins is always the book of the ruins. We live in a time when the book itself is in ruins, *eskhate biblos*. I want to read this poem as if the young poet knew that he both read and wrote the space he moved through. The young poet's war was about the end of hierarchy, hierarchy blasted away under the stolen aegis of a bent cross. The war was the second in a series—or the seven times seven hundredth, it does not matter, topologically number furls into metaphor—an infinitely divisible series of forays of movement into form. It is an ancient movement, this cross that both annuls and becomes form. Under its sign the idea of everything is no more privileged than the idea of any thing.

In the more recent interviews collected in *Conversations* Milosz speaks of his "incredible fastidiousness and [his] need for a strict hierarchy" (1987, 196). The old man the poet has become might not like this reading of the young poet. "Literature," he says, "is very hierarchical" (196). Yet what was cut was released into meaning, his poem knows this. "Our hope is in the historical," says Milosz, "because history as time, but time remembered, is something different from nature's time" (182). Yet nature's time, too, we are finding is time remembered. "History is important," says the social theorist Sandra Braman. "In a topologically mapped universe, the location of a point is less important than how it got there. . . . systems also have memories, and when they unravel will retreat by the routes along which they had previously travelled in the self-organizing phase" (1994, 13–14).[1] The knot springs to and from its form, rehearsing the manifold movements of its many tyings, knot on knot.

"Now walk carefully," the next movement of the poem begins. The poem slows into present tense, the meaning in movement is discovered. The light is mixed, not Lithuanian dusk but "a patch of blue" through a "ceiling caved in by a recent blast" (later we will see this is noon, as workmen sit and have their lunch upon a table made of books). The scene moves from the

---

1. I wish to note here my deep indebtedness to Sandra Braman's paper. For years a number of people I respect (especially John McDaid and David Porush) have tried both directly and indirectly to get me to understand second-order cybernetics, catastrophe theory, and autopoiesis. Perhaps all this tutoring finally took when I read Braman's paper, though I think it was something more than this. I not only owe my discussion of topology to her paper, I also owe my reading (chapter and verse, her citations mine) of Kwinter. In a similar acknowledgment in his *Electronic Word*, Richard Lanham suggests that he "does not come by [a certain quotation] honestly," a statement that (however tongue in academic cheek) puzzled me in a book that addresses the nature of electronic text, which is nothing if it is not that prickly flower, the clinging nettle. I came by my discussion of Kwinter quite honestly from Braman; I learned from her poetic and incisive writing and continue to do so.

metonymic to the metaphoric. The pages of books are "like fern-leaves hiding / a mouldy skeleton, or else fossils." Every discovered thing is discovered as written on, inscribed: the skeleton etched by mold, the fossils "whitened by the secrets of Jurassic shells" (just as this line itself is now written on by popular culture, a whole epoch crossed over into sign, tradestyle and copyright, the Jurassic marketed in every airport shop and shopping mall, cannot be retrieved to mean geologically until the market culture is done with it and still not then), the fossils in turn inscribed by tears of rust. The whole space in turn is written upon by the movement of these lines.

Yet inscription as always is questioned: it isn't clear in these ruins whether the shadow of a dead epoch is a living form. The poem presents the ambiguous figure of a scientist already like the guardian mole of Milosz's later poem, "A Poor Christian Looks at the Ghetto," who "distinguishes human ashes by their luminous vapor / The ashes of each man by a different part of the spectrum." In the scission of the caesura in the poem at hand (the English translation is Milosz with Robert Haas) the scientist's mind is split:

> A remnant life so ancient and unknown
> Compels a scientist, tilting a stone
> Into the light, to wonder. He can't know.

What at the start of the poem was the potential "you don't know yet" is now embodied, characterized and denied, "He can't know." The stone must be re-placed, tilted into light, before the scientist "looks again" and in this recurrence reads. The compulsion to wonder marks the figure of the scientist as no mere polar figure for the poet. A "rust of tears" simultaneously writes and erodes "chalk spirals," and the next caesura in the following lines ("Thus") is an engine, a compulsion to wonder:

> He looks again
> At chalk spirals eroded by the rain,
> The rust of tears. Thus, in a book picked up
> From the ruins, you see a world erupt.

The idea of everything is no more privileged than the idea of any one thing. Following the catastrophic eruption of world from book, "Green times of creatures tumbled to the vast / abyss and backward." The one who reads the book in the ruins replaces the one who reads the space of the ruin in the book we read. In its unraveling the erupted world retreats by routes along which it had previously traveled. Thus beyond the abyss of the poem the paleontologist's fossil gives way to the poet's imagined "earring fixed with trembling hand, pearl button / on a glove." Seen is scene.

In the course of what is seen the writer is replaced by the reader (the writer who will be). This is the claim of constructive hypertext, and by extension any system of electronic text, from hypertext to virtual reality to ubicomp. The fossil word, which on the computer screen is always tilted to the light and constantly replaced, again takes its place within the universe of the visible and the sensual. Print stays itself, electronic text replaces itself. With electronic text we are always painting, each screen washing away what was and replacing it with itself.[2] The shadow of each dead letter provides the living form of what replaces it.

The electronic text is a belief structure, and the workaday

---

2. Much of the discussion of electronic text in this vicinity (and indeed much of the language) has migrated from other texts and talks (especially "A Feel for Prose: Interstitial Links and the Contours of Hypertext"), many of them collected in Joyce 1995. In the introduction to that collection I argue as follows (aware that some might see such an argument for electronic text as an apologia for well-wrought sloth):

> What is . . . increasingly characteristic of the late age of print . . . is that before, during, and after they were talks or essays, these narratives were often e-mail . . . messages, hypertext "nodes," and other kinds of electronic text that, as will be seen, moved nomadically and iteratively from one talk to another, one draft to another, one occasion or perspective to another. The nomadic movement of ideas is made effortless by the electronic medium that makes it easy to cross borders (or erase them) with the swipe of a mouse, carrying as much of the world as you will on the etched arrow of light that makes up a cursor. At each crossing a world of possibility can be spewed out in whole or in kernel, like the cosmogonic dragon's teeth of myth. Each iteration "breathes life into a narrative of possibilities," as Jane Yellowlees Douglas says of hypertext fiction, so that in the "third or fourth encounter with the same place, the immediate encounter remains the same as the first, [but] what changes is [our] understanding." The text becomes a present tense palimpsest where what shines through are not past versions but potential, alternate views. (3)

reader is apt to believe that even the most awkward contemporary technology of literacy embodies the associational schema of the text that it presents. She sees herself there in its form as she finds it, and feels that the form trails behind her as she goes. Scene is seen: this is the shift from the book. Not that the storm of association around the book—the isobaric indices, the note cards, the pages "scattered like fern leaves," or the tropic marginalia—could not encompass the associations within it (they have and wonderfully, this is the history of the book as remembered time); but rather that the electronic form now embodies the same latitudes that once encompassed the book. The storm circles inward and disperses, belief structures saturate the electronic text, raining down like manna, driving skyward through us like the gravitron, sustaining and anchoring its continual replacement.

Likewise with electronic media the image again takes its place within the system of text, that is within narrative syntax, where, as Braman reminds us, "the location of a point is less important than how it got there." What's seen next: what's said next. The constructive hypertext is a version of what it is becoming, a structure for what does not yet exist. As such it is both the self-organizing phase of the reader who replaces the retreating writer, and the readable trace, time remembered in the unraveling retreat of this replacement. The one who will write will have to recall what the one before has written in such a way that the next one, the third self—you who write after us—may find the one finding the other first. Else what either of us have done will be lost to you and to remembered time; in its retreat the space of the story will neither mark the form of its making nor the making of its form. There will be no knot.

Seen is scene.[3] In the poem the erupting world "glitter[s] with its distant sleepy past." Seen through the poet's eyes, four chapters follow from the book in the ruins: three linked, a fourth marked by a scission and wounded. The first, as always, is light and lovers meeting.

3. My EESLA encyclopedia entry, "Hypertext and Hypermedia" (Joyce 1995), begins, "Hypertext is, before anything else, a visual form." Bolter 1991 is by far the best early introduction to electronic writing issues. It was followed by Landow (1997), which builds upon and extends Bolter's framework in important ways.

The lanterns have been lit. A first shiver
Passes over the instruments. The quadrille
Begins to curl, subdued by the rustle
Of big trees swaying in the formal park.
She slips outside, her shawl floating in the dark,
And meets him in a bower overgrown
With vines. They sit close on a bench of stone
And watch the lanterns glowing in the jasmine.

The big trees are twisted metal columns, we know this from the space where we entered; a bench of stone and glowing lanterns will become a table of heavy books dragged out by workmen in the light of a fire the sunlight kindles on a floor strewn with pages. You can't know this yet from the poem but you know it now from this particular reading. Or perhaps you already know this from the poem and mark it again in this reading. In hypertext each such point of a reading—what you know or don't know, tree or column, bench or table—potentially impinges upon another.

Yet we don't know where or in what form the world will erupt, whether for us, for the lovers, or for the poet who sees a book in the ruins. In "Treatise on Poetry" Milosz questions "whether Hegel's Spirit of History is the same spirit that rules the world of nature . . . in what respect history is a continuation of nature" (1987, 174).[4] For him, naturally, it is a question of being versus becoming, Heraclitus flows into Hegel, the confluence overflowing the Thomist channel.

Or so the poet says outside the poem, where he knows no other metric for nature or history. Inside the poem, however, there is, I think, the topological, a truly Heraclitian science.[5] "Space is best mapped topologically, rather than, as has been the Western habit, geometrically," writes Braman,

---

4. In following pages Czarnecka brings up the confluence of Hegel and Heraclitus (and the Thomist contrary) directly, asking whether the poet means "'Becoming' in the Heraclitian sense of the word as constant flow." To which Milosz replies, "It's pure Hegelianism" and then quotes a section from the poem that begins "O Antithesis which ripens into Thesis."

5. Kwinter suggests that "Catastrophe theory is a fundamentally Heraclitian 'science' in that it recognizes that all form is the result of strife and conflict" (1992, 60).

the latter is capable only of modelling linear movement or change . . . [describing] a system at a given moment only in terms of its earlier or later states, meaning geometric mapping can never describe the transformation of a system. Topology . . . describes qualitative transformations, including discontinuities so severe they transform the system itself. (1994, 362)

The jasmine light cannot last, scene is seen but sound is a singularity, a discontinuity, space furled in on itself. The next chapter of the book in the ruins is a stanza that won't stand still. Sounds come to the fore, the curl of quadrille and rustle of trees giving way to what you hear here:

Or here, this stanza: you hear a goose pen
Creak, the butterfly of an oil lamp
Flutters slowly over scrolls and parchment,
A crucifix, bronze busts. The lines complain
In plangent rhythms, that desire is vain.

The creaking goose pen, like the Escher hands that draw themselves (themselves now gone over into tradestyle and T-shirt, greeting card and wallpaper, an icon of recurrence), writes its own plangent lines, a genuine autopoiesis. In his "Treatise on Poetry" the butterfly serves Milosz as a sign of the power of recurrence; the boy in that poem (the section is a portrait of the artist as Lithuanian boy logician and poet) "looks upon the butterfly's colors with a wonder that is mute, formless, and hostile to art." Outside the poem Milosz (who is not afraid to say what something means outside the poem) says, "That means he admires the butterfly, but art which was supposed to be an incantation and break that cycle of recurrence . . . turns out not to have been very effective; the power of recurrence, of the natural order is very strong" (1987, 180). Whether this is so we cannot know, even holding it to the light, but it is clear that the butterfly in this poem flutters over icons of recurrence, from Horace's bronze bust to Jesus cross, from scroll to book, and alights on desire.

I once wrote about electronic text that "our desire is a criticism that lapses before the form and so won't let form return to transparency" (1995, 220). I am not wise enough to be able to say what that means exactly but I know I was trying to think about how to talk about the replacement of author in its double sense: the author moves to another place, the author is put in another place. "We need to surrender control and in that constant declination continually render control meaningless," I went on. The electronic text is such an oscillation, a strange clock that keeps track of space not time, or, if time, what Milosz in the "Treatise on Poetry" (1987) calls "Time lifted above time by time."[6]

Place to place within the electronic text, place itself is replaced in something like Braman's "qualitative transformations . . . discontinuities so severe they transform the system itself." So too, in the next chapter of the book of Milosz's poem, "here" replaces "here" in just such a transformation. The earlier line "Here, this stanza" becomes

> Here a city rises. In the market square
> Signboards clang, a stagecoach rumbles in to scare
> A flock of pigeons up.

Not just the birds (these pigeons like the hawks of the entrance to the poem), but everything (including time itself), is up in the air again. "Under the town clock / In the tavern, a hand pauses in the stock / Gesture of arrest" and meanwhile underneath (remember this is a clock in a poem in a chapter of a book in a ruin: remember, remember) meanwhile life goes on, "meanwhile workers walk / Home from the textile mill, townsfolk talk." The scene here is another metric, the human drama seen as a mechanical clock, and though, as Jay Bolter reminds us in *Turing's Man*, every such clock is a miniature universe, a recurrence machine, it cannot serve as a measure of mind. "The course of the planets seems unchanging and free of interference, so it can easily be mimicked by a clock," says Bolter, adding in his typical, understated calm, "But the human mind seems more changeable . . . responding in a variety of ways to new circumstances" (1984, 29).

---

6. Milosz 1988, 86. With Poincaré, Kwinter points out, "Time . . . reappeared in the world as something real, as a destabilizing but creative milieu; it was seen to suffuse everything, to bear each thing along, generating it and degenerating it in the process" (1992, 52).

Now while the poem is stopped under the shadow of the out-stretched hand, let us use this moment of calm as a place to talk about contours and new circumstances of mind. Previously I have talked about the qualitative transformations in electronic texts in terms of contours, borrowing a geometric term for what I now think I have always really understood topologically, sensually (as a caress) and outside the linear. I meant how the thing (the other) for a long time (under, let's say, an outstretched hand) feels the same and yet changes, the shift of surface to surface within one surface that enacts the perception of flesh or the replacement of electronic text. "If there is a name *surface*," asks the poet Erin Mouré, "then what else is there? is what is 'different' from the surface *depth* or is it *another surface?*"[7]

Contour, in my sense, is one expression of the perceptible form of a constantly changing text, made by any of its readers or writers at a given point in its reading or writing. Its constituent elements include the current state of the text at hand, the perceived intentions and interactions of previous writers and readers that led to the text at hand, and those interactions with the text that the current reader or writer sees as leading from it. Contours are represented by the current reader or writer as a narrative. They are communicated in a set of operations upon the current text that have the effect of transforming that text. Contours are discovered sensually, and most often they are read in the visual form of the verbal, graphical, or moving text. These visual forms may include the apparent content of the text at hand; its explicit and available design; or implicit and dynamic designs that the current reader or writer perceives either as patterns, juxtapositions, or recurrences within the text or as abstractions situated outside the text.

Topology is sometimes called "rubber sheet geometry," we can think perhaps of the supple and gelatinous movement of a jellyfish to animate it. In trying to formulate this provisional

7. In *Furious* (Mouré 1988) the collection of poems is followed by a section titled "The Acts," which is both an extraordinary poetic manifesto ("it's the way people use language makes me furious") and also a protohypertext linked to various poems by footnotes. The quotation here is from this linked section. Mouré's poetry seems to me hypertextual, especially the poems in *WSW (West South West)* (Mouré 1989).

definition for contour, I meant to recover this surface to surface and transparent shift that Mouré suggests in lieu of depth as a name for difference. Narrative syntax, like Brancusi's curve, is a syntax of merging and emerging surfaces. Yet even now the whole thing gets caught up and sticks upon a barb, the so-called given point in the claim that contour is "one expression of the perceptible form of a constantly changing text . . . at a given point in its reading or writing."[8] A point at first it seems must be a metric, a measure of a state and not the transformation of a system. The poem senses this, and won't wait much longer for us for tarry in the talk of townsfolk or the calm of new circumstance. We will have to put off this point of transformation, time moves on.

> and the hand moves now to evoke
> The fire of justice, a world gone up in smoke,
> The voice quavering with the revenge of ages.

So ends the third of three linked chapters of the book read in the ruins in the poem "A Book in the Ruins." *Eskhate biblos.* Doom descends in the figure of a man in a tavern whose hands show the final hour. Does this arm's span retrace Leonardo man as "measure of everything"? Everything seen now smoke, all sounds a quaver.

We are talking about transformations. In his extraordinary, topologically based meditation upon Boccioni, Sanford Kwinter describes how "in topological manifolds the characteristics of a given mapping are not determined by the quantitative subspace

8. Caught upon such a point, one naturally thinks of Derrida's meditation on the question of style:

> It is always a question of a pointed object. Sometimes only a pen, but just as well a stylet, or even a dagger. With their help, to be sure, we can resolutely attack all that philosophy calls forth under the name of matter or matrix, so as to stave a mark in it, leave an impression or form; but these implements also help us to repel a threatening force, to keep it at bay, to repress and guard against it—all the while bending back or doubling up, in flight, keeping us hidden or veiled . . .
> Style will jut out then, like a spur, like the spur of an old sailing vessel: like the rostrum, the prong that goes out in front to break the attack and cleave open the opposing surface. Or again, always in the nautical sense, like the point of a rock that is also called a spur and that "breaks up waves at the entrance to a harbour." (1985, 7)

(the grid) below it" (1992, 58),[9] which is, perhaps, to say that jellyfish's course is not mapped by the seafloor, nor by a world gone up in smoke, nor by an arm's span or a stanza. Instead Kwinter suggests that the characteristics of the manifold are mapped "in specific 'singularities' of the flow space of which it itself is part." The jellyfish is a form of water, we might say. "These singularities," according to Kwinter, "represent critical value or qualitative features that arise at different points within the system depending on what the system is actually doing at a given moment or place." Scene is seen, the movement of the map makes the mark. Which brings us to the point, the barb of metric, where we left off before the fall of the hand of doom.

"Singularities," says Kwinter, "designate points in any continuous process (if one accepts the dictum that time is real, then every point in the universe can be said to be continually mapped onto itself)" (1992, 58). Print (we remind ourselves) maps itself against the geometric and so stays itself; electronic text, conversely and topologically, replaces itself. What happens at the point of a singularity, says Kwinter, is "that a merely quantitative or linear development suddenly results in the appearance of a 'quality' (that is, a diffeomorphism eventually arises and a point suddenly fails to map onto itself)." Contours, I have said, are discovered sensually and they most often are read in the visual form. "A singularity in a complex flow is what makes a rainbow appear in the mist," says Kwinter (1992, 58). "So the world seems to drift from these pages," the poem goes on,

> Like the mist clearing on a field at dawn.
> Only when two times, two forms are drawn
> Together and their legibility

9. I wish to note again that the quotations from Kwinter 1992 woven through this and the following paragraph I first found cited so in Braman 1994.

Lanham (1993) also goes back to the Futurists. His essay "Digital Rhetoric and the Digital Arts" (29–52) launches itself Marinetti's *Futurist Manifesto* (and a less well known tract *La cinematografia futurista* from which he translates the following malediction: "The book, the most traditional means of preserving and communicating thought, has been for a long time destined to disappear, just like cathedrals, walled battlements, museums, and the ideal of pacifism" [31]) and works his way in almost effortless, and copiously illustrated, fashion to a discussion of why electronic text is like Christo's *Running Fence*.

Disturbed, do you see that immortality
Is not very different from the present
And is for its sake.

In an instant—the interstice of line to line in a poem, the turn of page in a book in the ruins—the smoke of a world gone up becomes a mist, last hours become the awkward light of dawn. We are back to the boy and the butterfly but now the problem of recurrence maps immortality on the present. Twin time is twined, and so less lonely.[10] The search is for the diffeomorphism: where the legibility of two drawn on two is derailed; where immortality differs, though not very much, from now; and where therefore the idea of everything is no more privileged than the idea of any thing.

When I was a boy, I spoke like a boy, St. Paul reminds us. Previously I tried to speak of the diffeomorphism of electronic text in terms of coextensivity and depth, where coextensivity is our ability to reach any other contour from some point of a hypertext, that is, the degree of impingement and dissolution among elements of a hypertext. Borrowing from Deleuze and Guattari (1983a), I suggested that coextensivity is the manner of being for space. As a boy in this I didn't know the names for the topological, and probably meant the jellyfish.

Depth was a more difficult thing, and shows itself as such in this poem. I characterized it then as the contour-to-contour inscriptions or links that precede, follow, or reside at any interstice along the current contour of a hypertext. Depth, I wanted to suggest—borrowing the second half of Deleuze and Guattari's formulation—was the manner of being in space, the capacity for replacement among elements of a hypertext. Our eyes, I said, read

10. This sentence plays on a literal translation of the Polish metaphor that I owe to personal communication with the Milosz scholar and translator Regina Grol-Prokopczyk, professor of comparative literature at Empire State College in Buffalo, New York. The verb *(splota)* has the sense of twisting strands of a cord around together (twining) until their legibility is blurred. The twining is also (re)enacted in the Polish, where eternity is said never to be lonely because it connects with the day in a double-stranded way: the phrase *(i po to)* having both the sense of "connected with day and for it(s sake)" as well as "connected with day just as well ('for that too')." That Milosz and Haas choose a metaphor of inscription for the English here seems telling.

depth as the dimension of absence or indefiniteness that opens to further discourse and other morphological forms of desire. Like any boy I probably did not know what I meant by depth or desire; I meant the flow space or the jellyfish as a form of water, but instead named petals on the starfish, points upon a grid.

The poem, however, has the word right. We are talking about form. Smoke into mist is a morphogenetic system, which, as Braman explains, "because [they] destroy form as well as create it . . . are also described [following Prigogine] as dissipative structures" (1994, 360). Braman points again to Kwinter and I look along with her (scene is seen):

> Catastrophe theory recognizes that every event (or form) enfolds . . . a multiplicity of forces and is the result of not one, but many different causes. . . . Any state of the system at which things are momentarily stable . . . represents a form. States and forms, then, are exactly the same thing. . . .
>
> In fact, forms represent nothing absolute, but rather *structurally stable moments* within a system's evolution . . . paradoxically [emerging at] a moment of structural *in*stability. (Kwinter 1992, 59).

The electronic text is a dissipative (belief) structure and the reader is apt to believe that its states and forms are exactly the same thing. The electronic text thus embodies a multiplicity of forces in the associational schema that it presents. Contours are not forms in the text, the author, or the reader, but rather those moments that express relationships among them in the form of the reader as writer. Originally I wanted to isolate coextensivity and depth in order to locate the place where the reader truly writes the text, the interstitial, which I now see as the "diffeomorphism [where] a point suddenly fails to map onto itself." The ability to perceive this point (we can call it form) makes genuine interactivity possible. It gives us some place both to move in and move the text; it thus replaces the author in the double sense of replacement.

We are talking about interaction. Replacement is the interaction that occurs when what Umberto Eco calls the "structural strategy" of an open work becomes available for reinscription.[11] For Eco, writing about interactive media in its earliest, serial epoch,

> the possibilities which the work's openness makes available always work within a given field of relations. . . . the *work in movement* is the possibility of numerous personal interventions, but it is not an amorphous invitation to indiscriminate participation. The invitation offers the performer the chance of oriented insertion into something which always remains the world intended by the author. (1989, 19)

Eco's author places us; "the author offers . . . the performer . . . a work to be completed. . . . It will not be a different work. . . . a form which is [the author's] form will have been organized, even though . . . assembled by an outside party in a particular way he could not have foreseen." Electronic text can never be completed; at best its closure maps point on point until time is linear and the text stays itself, becoming print. But when a point suddenly fails to map onto itself, the author is replaced. Replacement of the author turns performer to author. The world intended by the author is a place of encounter where we continually create the future as a dissipative structure: the chance of oriented insertion becomes the moment of structural instability, the interstitial link wherein we enact the replacement of one writing by another.

Seen through the poet's eyes, I said earlier, four chapters follow from the book in the ruins: we have seen three linked, now

11. For over a decade now (beginning with my first writing on these matters, "Selfish Interaction: Subversive Texts and the Multiple Novel" [Joyce 1995]), I have proposed in a number of talks and papers that Eco's open work and the notion of "oriented insertion" prefigure the transformation of reader to writer in constructive hypertext. Following one such talk—to the infamous, unruly, and influential CHUG group at Brown University—Robert Scholes's persistent and generous questions prodded me (rather unsuccessfully I must say, since I remained cheerfully utopian and dim about it all) to account for differences between the open text and the constructive hypertext. Even so I was (and am) grateful for his prodding, so much so I tried to account the differences more clearly here. The truth is that when I wrote "Selfish Interaction" I foresaw no difference between Eco's open work and the multiple fiction; but we had not built a hypertext system then. Nor had I experienced my students' apprehension of constructive hypertext or myself written hyperfiction.

comes a fourth, marked by a scission and wounded. The fourth as we reach it is, like the first, lovers meeting outside time. What cuts the caesura here, however, is shrapnel, an actual fragment.

> Disturbed, do you see that immortality
> Is not very different from the present
> And is for its sake. You pick a fragment
> Of grenade which pierced the body of a song
> On Daphnis and Chloe. And you long,
> Ruefully, to have a talk with her,
> As if it were what life prepared you for.

I want to speak carefully. I do not mean to suggest a cheap analogy, or to misappropriate the form of what might seem a fairly straightforward and romantic, even classic, poem; but what follows with Chloe is directly interactive. You can see it as a trope, an apostrophe; it is nonetheless the singularity that follows (here quite literally) a catastrophe, what life prepares us for. "How is it," the poet asks the woman in the fragment of the poem in a book in the ruin fragmented by a real grenade, "How is it, Chloe, that your pretty skirt / Is torn so badly by the winds that hurt / Real people . . . ?"

This is a question of forms beyond the book. Thus in the verse that follows, unscathed by what wounds her, Chloe—"who, in eternity, sing[s] / the hours, sun in [her] hair appearing / And disappearing"—here runs beyond the book and through the actual landscape, where real oak groves burn beside forests of machinery and concrete, pursued by the voice of the poet (a Daphnis we could say gone daffy with longing and grief):

> How is it that your breasts
> Are pierced by shrapnel, and the oak groves burn,
> While you, charmed, not caring at all, turn
> To run through forests of machinery and concrete
> And haunt us with the echoes of your feet?

In the old forms it is supposed to end in a wedding, this pastoral. The lovers are supposed to be replaced in their homes, restored to their parents, wedded and fed. The poet, sure enough, attempts it. "If there is such an eternity, lush / Though short-lived, that's enough," he says, but his own question cannot sustain the possibility of this form and muffles himself, "But how . . . hush!" the line ends.

Though not the poem. A marriage still took place here, albeit an ambiguous one, and so the banquet that follows is equally ambiguous. It is a marriage of life and forms beyond the book, writer and reader (the writer who will be). The groom makes a formal speech, it isn't clear to whom (it isn't clear now to whom he's been talking all along):

> We were predestined to live when the scene
> Grows dim and the outline of the Greek ruin
> Blackens the sky. It is noon, and wandering
> Through a dark building, you see workers sitting
> Down to a fire a narrow ray of sunlight
> Kindles on the floor. They have dragged out
> Heavy books and made a table of them
> And begun to cut their bread. In good time
> A tank will rattle past, a streetcar chime.

It isn't clear. Scene is seen. Noon is made dark by a ruin other than the one the poet is wandering through, a ruin in a romance outside time. Its darkness is illuminated by a fire kindled by sunlight other than that which, appearing and disappearing, marks the hours in the hair of the girl who lives beyond time. Perhaps the books are burning. Perhaps the books become an altar where the workmen cut their bread. Perhaps "in good time" the books (in time) lift time above time like sacramental bread. Who can know? Ask Milosz, perhaps he will tell you what this means. In Polish the poem ends on fingersnap, two words: *Tak Proste*, "that simple," which the Milosz-Haas translation folds into the ambiguity of "in good time."[12] Scene is seen but sound is a singularity, a discontinuity, space furled in on itself. Tank rattles, streetcar chimes, that simple.

---

12. Again I owe this insight to Regina Grol-Prokopczyk. She also points out that in the Polish the workmen place (rather than, as here, cut) the bread upon the table of books. The Milosz-Haas translation thus emphasizes the scissions of this poem of scissions.

I should end this there, borrowing against the rhythm and parabola of the poem, its apparent closure. I should end this here but I won't yet (though soon enough) because print stays but electronic text is spaces. Again I want to speak carefully. Again I want to speak of forms beyond the book. I want to leave off (not end) in another space, it too a library of sorts (though some would say a ruin), one in which I confess, though I am in some sense among its authors, I am as baffled and multiple and lost, as much replaced as Milosz is here in my reading of him.

Hotel MOO is an electronic space, a textual virtual reality (VR) if you will[13] (MOO stands for MUD Objected Oriented, and MUD in turn for Multi-User Dungeon or Dimension). There are literally hundreds of these virtual spaces throughout the world where anyone who has a computer connected to the internet can come and move through the written space in real time in the presence of others who write the space they read with their actions, their objects, and their interactions with each other on the screen. This particular virtual space is the creation of a Brown University student programmer and poet named Tom Meyer who built a structure (of words, everything I talk about from now on is made of words) upon the structure (the programming language for Lambda MOO) created in turn by Pavel Curtis of PARC upon the structure created by another student, Steven White, at the University of Waterloo some years ago. What Tom Meyer did was build a way that hypertext structures created in Storyspace (a program developed by Jay Bolter, John B. Smith, and myself) could become rooms in a MOO, their links doors or corridors (or windows) between them, each room a text to be read as well as written, and a place where at any time a reader could encounter another and a text change (seen is scene) before one's eyes. Hotel, the eponymous structure for this MOO, is a hypertext created (and continuing to be created) by Robert Coover, his students, and others over the past few years, first in a hypertext system

13. But you can really go there, if you know how, with Crotty and Coover and Chloe but not the extraordinarily valuable collection of texts by the women's hypertext collective, Hi-Pitched Voices, which was lost in a transfer of files from one MOO to another.

called Intermedia, now lost[14] (these genealogies, as we have been reminded by Foucault, themselves inhabitable structures and open to time), and then in Storyspace. Months since, other stories appended themselves like ramshackle additions to the lobby. A hypertext critical edition of David Blair's videotape *WAX: Or the Discovery of Television among the Bees* is, hivelike, being built there; a women's collaborative hypertext project Hi-Pitched Voices has linked stories under a yonic roof with corridors made of sentinas.

But the first time I was there, in the place where I sat in what is called RL, real life, it was two in the morning (though the MOO has no obvious time). Hotel MOO was new and there were only the original stories. I entered the lobby of the hotel by typing my character name and typing where I wished to go. I was alone there at that hour. I went to the Penthouse Bordello where a man and a woman both propositioned me (in fully developed, comically obscene dialogue). Robert Coover was also there (but, like the man and woman, in a story, a written self). I declined the advances of both man and woman, left the virtual Coover to eternally tell his life story, and took one of the "obvious exits," finding myself, surprisingly, delightedly, in a place between the walls of the hotel rooms, inhabited only by the rats (one of whom has written his or her own epic in another space) and the hotel engineer, a man named Crotty, whose story I unexpectedly found myself in (a story that was written, in the pseudo-Joycean protolingo of chaosmos, by me nearly a year before when I had visited Brown and was invited to put my mark in their hypertext). It was a shock to see myself there, and I was still in it when suddenly someone in real time asked me a question. She had come upon me without my knowing (you can find out who is in the MOO with you by typing a question, but when I'd come in there was no one; now too when you enter a room you can ask it to tell you how many authors it has had and what their names are).

"Hi, I'm new here," the person said. We will call her Chloe (though this is the truth and she did really appear there, the

14. See Landow 1997 for a rich account of this Atlantis.

reader as writer, typing somewhere elsewhere in the real of what once was my text). "Can you show me around?"

James Joyce could not be here with us today. Pavel Curtis wants to call the MOO a "social virtual reality" and in his current effort at PARC, Project Jupiter, a multimedia MOO, approach Joyce's ineluctable modality of the visible. "Every space within the virtual world would be a place where people could display text, bulletin board style," the *Wired* reporter (the ambiguity is their tradestyle) reports, "But people who were present in the same virtual room, no matter where they were in the physical world" ("Chloe," I say, though it could be Gerty McDowell), "could choose to be in audio or video contact." "Signatures of all things I am here to read, seaspawn and seawrack, the nearing tide, that rusty boot,"[15] that last is my cousin Stephen Daedalus speaking in a book, it too a library of sorts (though some would say a ruin). Actually he's writing in his mind there, moving along a strand and thinking of signs. What he reads he writes.

"In the course of what is seen the writer is replaced by the reader (the writer who will be)." I said this in another time (it was now, earlier in the space of this text), extending the claim of constructive hypertext to any system of electronic text, including virtual reality or (Russell's paradox notwithstanding) ubicomp. This claim is not merely meant to extend the dominion of the word within the universe of the visible and the sensual. It means also to point toward an inherent paradox, built into both the conception and the technology of virtual reality (both, of course, mapping onto themselves as one point), and which for now, and perhaps forever, makes it an alchemy. (Though there will ever be alchemists, a scientist in Texas even now attempts to work mercury to gold[16] and in a recent book Michael Heim suggests that "ultimate promise of VR"—"the Holy Grail" he says later—"may be to transform, to redeem our awareness of reality—something that the highest art has attempted to do" [1993, 124].)

More calmly Jay Bolter claims that in VR "the viewer gains in a mathematical sense more degrees of freedom. Like hypertext,

15. This appears, one hopes it goes without saying, in James Joyce's *Ulysses,* a different family of Joyce in every way (he, the master) from me.
16. See Mangan 1994 for the (fleet-footed) frontier details.

virtual reality invites the user to occupy as many perspectives as possible" (personal communication, 1994). Yet VR in some sense diminishes whatever the mathematical gain of degrees of freedom by insisting on the unification of all these perspectives within self-generating computer spaces meant to reproduce the illusory seamlessness of the aural, visual, and psychological world as thing. Freedom in my construction comes from replacement, the ability to scan and skip. VR attempts to subsume the point where point fails to map into the seamless. Against the backdrop of VR's infantile seamlessness, on the hierarchical stage set of the sensorium, the old drama plays, meaning by gaps (Lacan) or defamiliarization (Shklovsky 1990), by betweenus (Cixous 1991), by situated knowledges (Haraway 1991), or by my interstitial.[17]

The interdeterminability of points of perception[18] argues against a virtual reality that depends, as most do, upon successive disclosures of self-generating spaces. The last time I was in actual virtual reality,[19] no sooner had I donned the helmet than I went running full-speed for the edges of the representation, boundary testing, bursting through, blowing away the whole wire-frame world into a landscape of countless, brilliant ruby dice, each spin-

---

17. Lacan, like Foucault earlier, is cited from the ether—while the unseen and surrounding Chaos is, of course, Nietzsche. Benjamin Sher, making the stone stony, wants to translate Shklovsky's neologism (Russian: *ostraniene*) with the neologism *enstrangement,* though no one much wants to follow, including Gerald Bruns who writes the introduction. (As an aside, Derrida in the process of being called upon to defend, among other things, the study of his own death in the *New York Times Magazine* interview in January 1994, reports that the best translation is, as one might have guessed, *deconstruction.*)

18. I first used this neologism in a very similar sentence in "What Happens as We Go? Hypertext Contour, Interactive Cinema, Virtual Reality, and the Interstitial Arts of Jeffrey Shaw and Grahame Weinbren" (Joyce 1995). Perhaps I had in mind Haraway's "Situated Knowledges":

All these pictures of the world should not be allegories of infinite mobility and interchangeability, but of elaborate specificity and difference and the loving care people might take to learn how to see faithfully from another's point of view, even when the other is our own machine. (1991, 191)

19. At Andy Van Dam's Brown University Computer Graphics Group, which is managed by a wonderful friend of artists, Daniel Robbins, who was very tolerant, and even a little amused by my cheap trick of ruby dice, which I must emphasize did not in any way really "break" their program but rather simply went outside the events it was prepared to handle visually.

ning letter or numeral a particle of the code, each an error exception.

To be sure before long there will be (there already are) virtual worlds that can contain us wherever we run. But they too will be a structure of words, everything we will see from now on is made of words: scene is seen, text is useless, thus its use. On Vassar farm in virtual space or on a Dublin strand a woman walks toward me and passes, for all I know wordlessly, yet I believe that she too creates the space she walks through and that it is somehow different from mine. That is, there is no point from which to see, even in three-space, each new point in its own multiple perspectives. Gridded VR, driven by vision, misses the point that fails to map upon itself, the new thing, Milosz's barrier at the entrance, or the gate when (nor where) you go in.

Chapter 2

MOO or Mistakenness

When I first presented this essay as a talk at the MLA, I said: As is so often the case at the MLA this is not the talk I am giving. Or more accurately, what I said I was going to say was mistaken. This works out well since what I wish to talk about is the MOO or mistakenness. To do so let me tell you a story that at least two other times when I told it before was mistaken for another story.

I expected laughter at the initial line of my talk. There was none. (Actually I recall the faintest muffled guffaw at the rear of the room.) I am not going to tell the story that was mistaken for another now on the (perhaps mistaken) notion that, since I have spoken and then printed it elsewhere thrice now, *(a)* a reader might know of it; *(b)* I might better summarize and refer to it; *(c)* it had relevance only in the talk I gave (either the MLA talk or the talk where it first appeared, before appearing in print the second [but not the first] time, as part of the "End of the Book" conference in San Marino and thus the previous chapter here [perhaps making this a fourth appearance, if you count twice in the same essay]); *(d)* the inclusion of a reference to it in a paraphrased or quoted (it isn't clear which it is, given that in the first paragraph above I include the quote following a colon and without marks of [en]closure to suggest where "what I said" ends; so unclear in fact that I might still be saying it here; of course, italics might have helped [there] since when I last told the story in the journal where it appeared, though in another volume—or is it number?—likely, now that I think of it, both—I placed it in italics to make it clear it was quoted from both the previous talk and the publication leading from it, which publication would in fact follow the italicized quotation here, or—as a matter of fact—also the unitalicized [romanticized?] one above since both *Works&Days* instances appeared before the collection of the San Marino essays [note that I can assure you that this parenthetical tangle of clauses parses, though I wouldn't expect you to go to the trouble to do so]) statement above itself now stands in fact metonymically as the story (i.e., the one mistaken for another) to which it refers; or *(e)* its ellipsis here would serve the intent of an essay on verbal mistakenness and virtuality.

In the case of either the initially mistakenly playful conundrum of the public talk or the tortured (and equally laughless [though deeply parenthetical]) paragraph above, there is a calculation that the verbal construction will interest an unknown inhabitant. (We could call this person the reader, in which case I would be talking about you. Or you could say, "I suppose this sort of indulgence interests *him*," in which case you would be talking about me. In a MOO, of course, you *could* say all this. And I could call you the reader and talk about or to you.) This is the (current at least) nature of the MOO, which (as surely others have said elsewhere, though certainly not quite in the conundrumming language that I use beyond this parenthesis; that last phrase a spatial certainty that became something more than a tired modernist textual trick [tic] since the onset of electronic word processing in which the linearity of the beyondparenthesis can be filled with mutable, movable text already, as what follows here is, having once appeared only a chapter ago in another setting, in hand) is an electronic space, a textual virtual reality, a verbal construction meant to interest an unknown inhabitant, where anyone who has a computer connected to a network can come and move through the written space in real time in the virtual presence of distant others, joining together to write the space they read with their actions, their objects, and their interactions with each other on the screen (doing so in a way you can't here, or can you? It turns out that the whole issue of the journal where this first appeared was to have been itself edited through a "Listserv discussion . . . to make this issue a more collaborative and dialogic project, and for each of us to comment on others' perspectives and analyses," and thus you could have, or may have; or perhaps as Jay Bolter says in the [modernist tic tock] beginning of the hypertext version of *Writing Space:*

> *All readers should be aware that anything in the text may have been added by someone other than the original author.* But, of course,

this caveat applies in a Borgesian or Calvinolike way to the previous sentence as well). (Bolter 1991, n.p.)

The assumptions I make, including the faux self-reflexivity of the properly postmodern, seemingly antic, surely self-conscious, all too serious self I construct in the paragraphs above, also of course assume a role: *The Autobiography of Alice B. Toklas,* the "Autobiography" of John Barth's Funhouse; Roderigo S. M. writing as/and Clarice Lispector at *The Hour of the Star;* the literary estate of Arthur "Buddy" Newkirk in John McDaid's hyperfiction *Uncle Buddy's Phantom Funhouse;* Judy Malloy's storyteller in the Brown Kitchen at Lambda MOO; or my student Kate W. as Kate (rather than her accustomed character Fucky Mouse; which of course isn't true, since it would distress her mother, or anyone's [father also] to say so) on Vassar MOO when Mark Taylor comes to visit and "talk" about *Imagologies.*

The MOO is not about the level of (literary) wordplay that I am indulging in here (and which lead you earlier to say, "I suppose this sort of indulgence interests *him*"). Or at least that (the nonaboutness and the indulgence both) is the claim of Erik Davis:

> MUDs take on a more fantastic light when they're seen not as baby steps on the golden road of total immersive VR, but as the apotheosis of writing. Most computer literate highbrows have pegged hypertext—the permutation of narrative as nonlinear webs of linked textual objects that can be read in countless paths as the likely site for the emergence of computer "literature." But MUDs create nonlinear texts in many ways more marvelous than the precious literary experiments loved by Robert Coover. MUDs make text interactive, spontaneous, and collaborative; writers cobble together a collective hallucination (the rooms, object, and characters), breed narratives of love and war, and jam like improv poets with their chat. (Davis 1994, 42)

Even so I will discuss the MOO as hypertext. (Did one of us forget to say that a MOO is a *MUD Object Oriented?*) Partly this is the character I assume in this verbal construction, mostly it is because the MOO is nothing if it is not hypertextual, that is, as

I've defined it elsewhere (and now also here it seems as well), reading and writing in an order you choose where the choices you make change the nature of what you read or write.

I was in what is commonly known as Robert Coover's Hotel MOO when the story (i.e., the one mistaken for another; the phrase on the left side of this semicolon of course having been "grabbed" from where it was copied above and pasted here on the right in something of a visual approximation of, i.e., a virtual, Homeric epithet, e.g., rosy-fingered mistakenness), which I won't (I repeat) tell here, but which when I told it at least three other times before (including the MLA) was mistaken for another story, took place. ("Took place" isn't merely a phrase, of course, when regarding the MOO. There temporal language is mistaken for spatial, if only because they appear so. What is said there is printed there, though not for long ["print stays itself, electronic text replaces itself" as I've said in chapter 1 and elsewhere], unless one longs to log it. In MOOs language takes place; words take the place of places.)

It is a mistake to think of Hotel MOO as Robert Coover's MOO. In fact (instead) Brown University student programmer and poet Tom Meyer devised a way that hypertexts created in Storyspace could become rooms in a MOO. Each room became an imaginary wall of text to be read as well as written upon and thus a place where at any time a reader could encounter another and a text change before one's eyes. Hotel, the eponymous structure for this MOO, was a collaborative hypertext fiction created over the past few years by Robert Coover, his students, and others working first in a hypertext system called Intermedia, now lost, and then in Storyspace. So it wasn't Robert Coover's MOO in that sense. But Coover also (and always) was wary of the MOO, and was reported as saying of the reading experience of Hotel MOO in *Lingua Franca:* that it was "very diffuse, and seems totally undirected, and is finally a kind of bore to move around in" (Brennehum 1994, 31).

And yet Coover's mistakenness here is exactly the point at hand (so to speak): the confusion of the space for the occasion, I will suggest below (this deferral of time—again!—[as I noted above in discussing the modernist tic tock or the temporal play of what comes next on the other side of a semicolon or parenthetical hoop (or bracket)] is itself utterly germane to the MOO experience, which always defers [to] time, to what I will call below, I know now, writing in the spatial loopback of the temporal space of electronic writing, itself now looped into a double knot, a catastrophic singularity, in which both this interrupted sentence and this long parenthetical that interrupted it, are headed now to the same merged point) is the essence of MOO experience. And so when Coover goes on to say that "Maybe thirty years from now, it will be like the Mickey Mouse watch . . . [and perhaps] retain a kind of aura of its time that people will find compelling," he too, however inadvertently, signals an awareness of the deferral of time and its promises, whether in narrative, virtual experience, or history and/as popular culture. Coover may be right in saying it (the space) is a bore to move around in; however he may be mistaken in saying it (the occasion) is a bore to move around in. In any case, his mistakenness (or is it mine?) raises the question at (Mickey's-Mouse's not my, not Fucky Mouse's either) hand.

For in the story I am not telling I had, in the early days of Hotel MOO, and late on a winter's night, just left Robert Coover (who wasn't Robert Coover but a so-called bot, i.e., [ro]bot, in the MOO, a Coover who constantly tells his own life story; something a literary critic [Larry McCaffery, for instance, though it would be mistaken to think that he did] could argue that the real Robert Coover does, but I never would, i.e., argue that he does) and had taken (i.e., I had, not the real or virtual Coover) one of what the screen said were the "obvious exits," finding myself—surprisingly, delightedly—in a place between the walls of the hotel rooms inhabited only by the rats (one of whom has written his or her own epic in another space) and by the hotel engineer, a man named Crotty, whose story I unexpectedly found myself within (a story that had been written by me nearly a year before when I had visited Brown and was invited to put my mark in their hypertext and that I've told, in very nearly this same form, a mere chapter ago if you've been reading along *[a]* and bothered to follow the whole confusing mock epic *[b]* and who's *[c]* when she's at home?). It was a shock to see myself there and I was still in it when suddenly someone in real time asked me a question. We will call her Chloe. (The reason for this is that she exists in another text, or three of them; the characters in a MOO also exist in another text, or two, or more: from object, to description, to the interactive session.) She had come upon me without my knowing (you can find out who is in any MOO with you by typing a question, but when I'd come in there was no one; now too when you enter a room in Hotel MOO you can ask it to tell you how many authors it has had [prefacing your question with the iconographic @ sign, the ideogram of the alphaeye or the confusion of the space for the occasion] and what their names are, and how many people have followed its obvious exits).

"Hi, I'm new here," Chloe said, typing somewhere elsewhere in the real, "Can you show me around?"

Her greeting was a typed line in a stream of other texts, her gender perhaps only a marker, her location a set of network paths; she was the reader as writer.

The name of this story I am not telling here (token parentheses) was once the interdeterminability of points of perception or a short stay in a small hotel in cyberspace. The point of the story, in my mind at least, is (was) to ask in what sense I could reply to Chloe's question. What would it mean if I answered this apparent woman as the ambiguously named character, Joyce, the name I had been given on the MOO (a name that, I might note, during the official opening reception of Hotel MOO had subjected me to constant harassment by someone whose words claimed to be repeatedly mounting and humping me as I chatted with network friends and my Vassar chair, a woman of some dignity). What would it mean if I answered as the hotel engineer, Crotty, or at least an acquaintance of his within this hivelike, serial story. What would it mean if I answered as the author of the text that formed the walls of the room in which our scrolling words, like winter exhalations, formed the visible but dissipating air. In my mind the point of the story was that each of my possible responses would have represented an intrusion of my dubious authority upon the late-night traveler. Yet twice when I told the story others mistook it as a story of her intrusion upon me.

(First telling.) Some years ago at the NEA's "Arts in the Twenty-first Century" conference I offered the story as a literary equivalent of my fellow panelist Alan Lomax's "Global Jukebox," an instance of networked collaboration suggesting the inevitable blurring of boundaries between the writer who was and the writer who will be. Yet after the talk a woman from Hawaii, whose name I have sadly and tellingly forgotten, sought me out, worried about what I had said. A novelist interested in cyberspace, she saw the story as one of intrusion, the loss of the authorial space, rather than one of collaboration.

As I am by no means the first to note, talk of how collaboration dislodges individual authorship seems especially problematic to many observers and seemed especially so that year at a time when three women of color simultaneously had books on the *New York Times* best-seller list. In such a light, hypertext, the web, or MOO space threaten to virtualize cultural white flight; technoculture becomes a strip mall, relocating and decentering at exactly the moment that previously marginal voices finally inhabit the center.

Even so I think she was mistaken. Technoculture broadens the center to a smear of light, authorial space is spread across its surface as if the Milky Way or peanut butter. (Is this the wrong language for something so serious? Am I mistaken in my register? The MOO too (and how are you?) alters the registration of serious discourse toward the antic, Barthes [or barters] bliss.) By distributing the center at a time when the margin(alized) have reached it, the newly centered margin is amplified. This is a property of waves. Or a hydraulic theory.

(Telling two.) At the "End of the Book" conference I again offered the story, this time as an instance of the literal replacement of the author within hypertext, the author reading the reader's writing of the text. I wanted to suggest that in MOOspace as in hypertext the author's voice remains (or sustains) a place of encounter where we continually create the future. Eco however heard in my MOO story an instance of one tier of what he proposed as a three-tiered taxonomy of hypertexts: hypertext as system, hypertext as artwork, and hypertext as what he (flatteringly for this occasion) called "Joyce's jam session" (restated in Eco 1996, 303). The last ironically inverted Erik Davis's already cited dismissal of hypertext as mistakenness in the *Village Voice:* "Most computer literate highbrows have pegged hypertext . . . as the likely site for the emergence of computer 'literature.' But MUDs create nonlinear texts . . . [where] writers cobble together a collective hallucination . . . and jam like improv poets with their chat."

On the MOO we are always mistaken. Mistaken for someone else (much as Joyce was at the Hotel MOO reception) or mistaking the form of the interaction as Eco and the Hawaiian woman did. Why this mistakenness? In Hotel MOO the rooms mean to be both the expression and occasion of their interactions. The room is what you read and where you write (sometimes writing what you read and where you write the next time). The Hawaiian novelist did not understand that the textual walls served as occasions for the encounters within them. Eco saw the encounter within the walls as isolated from their texts surrounding them.

Each was looking elsewhere. This is inherent in the MOO paradigm with its emphasis on being elsewhere, what we might call its exoskeletal virtuality (e.g., the MOO command @describe is outside the body; @page is outside the space, @who discloses that everyone is elsewhere). In the MOO attention is always elsewhere, it is a distraction, a disposition of self, the confusion of the space for the occasion. Any MOO experience is filled with the comings and goings of characters who herald their own movements with elaborate envoi that recall Groucho Marx's processional, "Hello, I must be going."

It is not surprising that the MOO is a locale of mistakennesses. There is a general feeling that the MOO is a mistake of technological history, a developmental lacuna, a place marker, an interregnum in the immanent hegemony of the postalphabetic image. Seen as such, MOOs are founded on a temporary appropriation of virtual space by the word for want of bandwidth for the image. There is everywhere a mixed feeling of apprehension and anticipation—among humanists as much as among technoids—that the MOO moment is temporary and that soon the image will either rob us of the power—or relieve us of the burden—of language.

The MOO is not so much an immersive technology as an expressive (in the root sense, the moving outward, shucking off the body). It is the ultimate (literally) WYSIWYG world (every-body's always awaiting a new planet), the last gasp of the alpha-betic (the MOO is WYSIWYS, what you see you say, where see and say oscillate). We didn't suspect that at this late date the alphabet (e.g., viz. Fenollosa, i.e., FYI, WYSIWYG is an ideogram, IMHO) could be a visual medium; we were mistaken if we thought this reprieve meant that would mean much. As Marky Mark never said to Fucky Mouse:

> In the media, one liners are everything. Impressions are every-thing. Style, personality and timing are everything. There is no possibility, and this cannot be emphasized too much, of ruling out the scholar's nightmare of ambiguity and, even more shocking, radical, outraged, emotionally charged misunder-standing. For those who still believe in the dream of transpar-ent intersubjectivity or an ideal speech community of the experts who trade clear and distinct ideas, essences and con-cepts, misunderstanding constitutes an abiding fear. But mis-understanding can release energy. The law of the media is the law of dirty hands: you cannot be understood if you are not misunderstood. (Taylor and Saarinen 1994, 5)

Maybe we (were) never meant to mean much. Which is to say (what it says) that the law of dirty hands is likewise the law of Fenollosa's ideogram (as EP had it: language that carries its history on its back). We mean by being (in words or elsewise [elfwise]). The MOO as a version of (or displacement of) hypertext depends on a misunderstanding of the degree of control we exercise in the release of energy (bricolage or any jazz). We mistake what's next for what's there and vice versa. The MOO is nothing if it is not the verbal representation of the hope of the rounded corner. Our sense of what is simultaneously both next and there is what I mean by hypertext contour. If as Erik Davis suggests (echoing [how consciously?] Gibson's definition of the matrix) the MOO is "a collective hallucination," then our shared sense of contour is what induces it. We induce (and seduce—perhaps even produce) the reader to join with us in shapeful behavior, recognizing emerging contour within the disclosures of the text. The contour is the figure of changing change, or how meaning is made. This is true of hypertext and the MOO, ring-a-rosy and dervish dance. Contour is one expression of the perceptible form of a constantly changing text, made by any of its readers or writers at a given point in its reading or writing. (Its figure for me, however, is the body, how under caress the curve is always another space and yet the curve of the caress, i.e., the nature of changing change.) Like other electronic texts the MOO is written as (and where) it is read. Yet sometimes we feel trapped and, as Marie-Laure Ryan suggests, we wish we could know what we will read before we have written it:

> The best way to prevent this feeling of entrapment, it seems to me, would be to make the results of choices reasonably predictable, as they should be in simulative VR, so that the reader would learn the laws of the maze and become an expert at finding his way even in new territory. (1994, n.p.)

This is the essence of my sense of hypertext contour, an expertness in the new. How we write the curve we read in the caress. In some sense, of course, the new is all we are ever expert in. We wish for ever we knew new forever. This wish for new knowing is mistaken, it too as much a trap as chance. The MOO, like hypertext, is founded on mistakenness, in this case what Terry Harpold terms (regarding hypertext fiction) "a duplicitous faith in the principle of chance, because chance is always assumed to be meaningful, which is the same thing as assuming there is no real chance" (1994, 193). Ryan likewise sees (seize) this:

> But if the reader becomes an expert at running the maze, he may become immersed in a specific storyline and forget or deliberately avoid all other possibilities. He would then revert to a linear mode of reading and sacrifice the freedom of interactivity. (1994, n.p.)

As Calvino pointed out (not so) long ago, "once we have dismantled and reassembled the process of literary composition, the decisive moment of literary life will be that of reading" (1986, 15).

Alison Sainsbury talks about reading hypertext in terms of "all the joyous experiments in which texts are freed from the tyranny of inheritance and primogeniture, texts in which circulation for the pleasure of circulation does not come as a matter of course, but courses through the text" (1995a). What courses through the MOO text is how we make meaning, as if a caress, or meet it around the corner, as if in a caress a curve.

"And if what I say about circulatory systems in a hypertext is even remotely so," says Sainsbury, "where, one might wonder, does oxygen come from? How does the breath enter? What 'system of exchange'? I think it comes from the reader, from the constant repositioning of text/reader, of what Carolyn Guyer has called the buzzdaze."

That is, she sees the reader as the tissue of lung, the surface to surface that the poet Erin Mouré talks about in lieu of the old surface-depth hierarchy. Surface to surface, face-to-face, we speak as Mouré says in the "hidden tensity of the verbs without tense. Because the past tense exists IN us speaking, or is not anywhere" (1988, 98). In the MOO this hidden tensity suffuses the diffuse boredom Coover mistakes for the space we move in. Likewise in Guyer's sense of the buzzdaze interactive vision moves through confusion to coherence and then back; thus the buzzdaze is the space between the oscillating poles, ultimately in us. "To make story, to write ourselves," we must Guyer says realize that "creativity is more about elaboration than fabrication, so [too] hypertext is more a verb than a noun, more about the *flux* of making, it is a *reforming* rather than a form" (1992a, 6; see also Guyer 1996b).

What does the MOO (re)form? By looking back upon itself both literally and self-reflexively, MOO language seems an instance of language longing to become and even transcend image. This mistaking of language for image is the star that has led (at least) Three Wise (Live White) Men (Bolter, Lanham, and W. J. T. Mitchell) to suggest the return of the rhetorical figure of *ekphrasis,* language vying with image. Lately the MOO has been dying on the vying as during this past year or so MOOs (and maillists and gophers and donner and blitzen) suddenly became if not deserted then quieted by a mass exodus toward ekphastic texts on

the world wide web, solitary instants (the web as premature ejac-
ulation?) displayed in a mosaic netscape and saved as beadwork
bookmark or hotlist wampum.

It is of course mistaken to claim that the web swallowed the
MOO. As any- or everyone will rush to point out, the web isn't a
thing as much as a writing of a thing (tag UR it). Thus it wasn't
long (it wasn't longing either) before the MOO became inscribed
as another Universal, its locutions located in a locator, in the
heart of the heart of the tel(tale)net, swallowed up in the maw
(ma) of the voracious web, the unidirectional link to every new
known thing, the one-way ticket to oneself. The web so seen was
MOOspace become solitary and asynchronous, with objects but
without characters, its glittery beads lacking the present-tense
link of what I call interdeterminability: how every other writes
what we write without and yet within our knowing.

Even so the web simply represents (despite the wariness many
of us have felt about attributing hypertextuality to everything)
the essentially hypertextual nature of computing in general (and
the net in particular) in which there has always been, as there is
in the MOO, an underlying representation of the availability of
processes, objects, and representations across the visible surface of
the interface. The great leap toward the web is more a recognition
than an original creation of virtual space and a locale for its
embodiment in our individual actions (including mental ones)
not unlike the embodiment of our multiple actions in the MOO.
(Though increasingly websites and browsers seek to embody the
multiple, whether in the shared inscription of the graffiti wall or
the shared journey of the Sociable Web.)

MOOs are places where we can, in Donna Haraway's words,
"construct and join rational conversations and fantastic imagin-
ings that can change history" (1991, 193). In MOOspace this construction and thus the history take the form of things, computational objects, scrolling text streams, the evidence of others'
perceptions of us. Any interactive installation is from bottom to
top a polymorphic weave of its own becoming, what in defining
constructive hypertext I have called a structure for what does not
yet exist. Its network of signs creates itself as a space within which
there is always written a layered history, a core sample from RL to

VR, from to the cyclic, constant replacement of the text on the screen to the turbulent flow of electrons (like the taste of peach) across the rilled chip beneath.

Life in the middle of things takes place in the present tense places of the heart. *In media res* is, of course, an ancient term for poetics and a description for a sense of life that I take to be characteristic of women's thought. To be in the middle of things rather than in opposition is what Haraway calls "knowledge tuned to resonance not dichotomy" (1991, 195).

For some time I have been trying to think of the situation of what we might call "realization" in cyberspace, that is, how the reader must make the text real in the context of Guyer's "complex mixtures of polar impulse . . . the buzzdaze of spacetime" (1992a, 4). The question at hand is what Sainsbury terms the "system of exchange," how do we breathe out and in? In this too, I have been mistaken. How could we not breathe (out or in). Erin Mouré says that

> even if you go into a cafe and tape somebody's normal conversations, it's full of nonsequiturs and little bits and pieces that are really discordant and even contradictory—and that to me is part of the nature of human thinking and of people's relating to each other. It only seems discordant because we've been trained to read or listen in a certain way, and the lyric poem, too, the narrative poem is constructed in order to coincide with the conventional notions of reading and, you know, who are those serving? The interests of newspaper advertisers. . . . The reader has to switch too, has to be prepared to say, "Okay there's more than one kind of reading." (1993, 38)

There is more than one kind of reading.

Interdeterminability is what happens as we go through the buzzdaze, each of us writing a mingled history of the self as sign and desire. In hypertext the web of possibilities is so richly layered and distributed that what is around the corner is not merely also simultaneous there but also somehow more strongly bound than what is here. Likewise in MOOspace every instant @join(s) the adjoining.

Like any virtual reality the MOO is founded on this mistaking. The fatal flaw of most VR design is to mistake our experience of the world for the world we experience. To think that the machine must represent the illusory seamlessness of the aural, visual, and psychological world is to misunderstand the undeniable and inevitable mediation that the presence of others, that is, the interdeterminability of points of perception, represents.

In the MOO we are mistaken if we think the story of what we see is the story of who we see or what they see. Yet who we see tells the story of what we see. A log of our encounter in a MOO session is, in some sense, impossible. Even were we to combine the logs that comprise our individual views, we would not have the text of around the corner, neither all the unseen choices nor the unknown encounters (even the ones we have not chosen in order to choose the ones we have). We mistake our experience of the world for the world we experience.

There is more than one kind of reading.

In its primitive, passing, and thankfully not yet seemingly seamless virtuality, MOOspace still allows us to see the folly of (even postalphabetic) VR design based upon the unification of perspectives within successively disclosed and self-generating computer spaces. Late at night in a small hotel in cyberspace I think a woman walks toward me and speaks. (This sentence echoes another you did not see in the story I have chosen not to tell here.) Perhaps I am mistaken but I think she is Pimoe, the Hawaiian shapeshifter. In any case I believe that she too creates both the space she walks through as well as the me she walks toward and that both are somehow different from mine. She asks me a question. Mistakenly I do not think I know the answer. I tell her a story. It is the story she tells me by her presence (a story not told here), the interdeterminability of points of perception. In it I begin by saying . . .

Chapter 3

New Stories for
New Readers

In November 1995, I was discussing N. Katherine Hayles's essay "Virtual Bodies and Flickering Signifiers" with my Vassar first-year students. We were talking in room 006, the computer classroom where IRL, or "in real life" as the computer jargon has it, a number of my students were actually physically seated. Others however were in their dorm rooms or the library. I myself was seated before a brimming ashtray of someone else's cigarette butts and the paleontological crusts of someone else's pizza at a computer in the basement room of an art museum in Hamburg, Germany, where I was attending the Interface3 symposium regarding the nature of the body and human community in cyberspace. The room 006 where my students and I met and talked was also in cyberspace, at Vassar MOO, the mirror world of a student-created, increasingly rich verbal representation of the Vassar campus as a textual virtual reality.

In the midst of our class discussion the computer in Hamburg beeped to tell me I had new mail on my account at Vassar. It was from Kate Hayles, the author of the piece we were discussing in the MOO class, firming up details for a talk she gave later that month as part of a Vassar Library lecture series regarding women and technologies. Somewhere in the shifting space of the infoscape, the present and the future, as well the present and the absent (or what we have come to call the real world and the virtual world), had briefly merged.

What is most amazing about these stories of technological presence and multiplicity is how used to them we are. A week or so after I returned from Germany I gave two workshops at the University of Missouri Institute for Instructional Technology

I wish to thank Sharon Dolente, Heather Malin, and Noah Pivnick, all Vassar College '96, for permission to quote from their theses and to recount our discussions.

(MUIIT) at the invitation of Eric Crump, a moving force within computers and writing community. I was seated in my real office at Vassar but speaking in the virtual MUIIT seminar room located IRL at ZooMOO on a computer in Missouri—"speaking" to conferees in three states. As yet another instance, in 1994 members of my hypertext rhetorics and poetics class got increasingly annoyed with the text about media philosophy that we were then reading, *Imagologies,* coauthored by Mark Taylor, a philosopher at Williams College. Without my knowing or doing, a woman in my class took it on herself to invite Taylor by email to meet with the class on Vassar MOO, and within two weeks he was "there" (or here), standing up to a barrage of questions from an energetic, skeptical, and informed community of his readers. I am quite certain that most of us knew the difference between meeting Taylor on the MOO and meeting him both IRL and FTF (which is computer jargon for "face-to-face"), but I am likewise certain that none of us thought we were merely continuing to read in this utterly textual encounter with him.

The computer, like our classrooms, becomes a theater of our desires as well as our differences. While as I have noted we are increasingly used to stories of technological presence and multiplicity, our profession as teachers and writers is predicated upon a necessary and creative scrutiny of the things we are used to, especially as they have to do with our understanding of differences: between the virtual and the embodied, between the lasting and the transient, between the rare delights of human community and presence and the universal promise of access and equality.

In coming to teach with and talk about electronic forms of writing and discourse we need to be aware of our desires and wary of what we are rapidly becoming used to in their representations. For what we are used to we too often become used by. In what follows here I want to discuss those representations and then suggest ways that we and our students can become unused to what merely passes as the new and so avoid becoming used by passing technologies. Finally I want to suggest a new voice for our interactions, one grounded in age-old values and a forgotten syntax, but with the promise to redeem us from the game of technology and into a real space for the play of self and community, a middle voice where we begin to see ourselves in where we are.

Increasingly where we are in on the web. The truth is that I don't enjoy the world wide web very much, except (and I don't mean this to be a punch line) when I am there for some reason or in the company of others, my students or my sons. Likewise I don't particularly find it a place as much as a utility, don't find it a medium as much as the virtual machine that people like Netscape's Marc Andreesen and Bill Gates of Microsoft have begun to acknowledge it as: the ur-computer that contains and enacts an increasingly dazzling, but often alienating, set of representations and practices. The web is a virtual machine in an actual machine, a shared and imaginary computer that contains virtual visions of actual us and that often disseminates others' actual plans for a virtual us.

The truth is that the web puts me at a loss and I do not exactly know why. Sometimes I fear mine may simply be the reaction of someone with a minor reputation elsewhere who feels himself passed by in this new set of practices. It is, ironically, the pure boundedness of the linked space that will distinguish my field, hyperfiction, in the age of the web. Thus I write a passing form in an uncertain medium. Passing since in hypertext the word is likely to have to renegotiate its relationship with images and audience, and uncertain because there is no guarantee that any of these works will survive the shift from virtual machine to machine as computational platforms change (in a process analogous to the shift from clay table to parchment).

I suppose my resistance to the web could be a hypertextier-than-thou reaction to the failure of the web to take the form I imagine it should have taken. Yet the fact is I don't know what form the web "should" take. In fact, I am quite fond of Rick Furuta and Cathy Marshall's (1996, 183) caution to the hypertext community that it is less important that we ask why the web has not paid heed to the decades of research into hypertext as it is to ask what makes the current forms so attractive.

And yet even as I say these things, I am suspect. I am aware of the telltale litany of "the truth is" and "in facts" that threads through what I've said already and that suggests that I may be deluding myself. The truth is likely something I cannot see and the facts somewhere beyond my saying. What we are used to we too often become used by, and so we must begin to see ourselves in where we are.

I am afraid that one of the things that makes the web in its current form so attractive is that so many of us are afraid and lonely and do not know what to think or who will hear us. The web does strike me as a lonely pursuit, something that douses the crispness of difference and community in a salsa of shifting screens. It encourages pecking orders and hierarchical thinking. Our culture has slipped the web over our heads like a lonely guy slips on a T-shirt from the Hard Rock Cafe. The web privileges the culture of brand names and corporate logos over the weave of our own multithreaded culture and history. Like the lonely guy in the Hard Rock T-shirt, we haven't anyplace to go. The old places seems either deserted or dull and the infoscape seems paradoxically crowded and lonely, already mapped by somebody else and often without a clear place for us. The web too often packages rather than represents the shape of our desires; and doesn't yet manage to show us what we really see there. We are caught between a Hard Rock and a *Melrose Place.*

It is no revelation to suggest that the web is more hierarchical than hypertextual. That is, the web is paradoxically and inherently (if sometimes innocently) hierarchical as a cost of its platform-independence and server-based links. If I move from your page to a certain site you point to, I am obliged to come back to your page if I wish to visit another of the sites you point to. In the process of doing so I begin to construct a contour that represents at very least the momentary ranking of my own interests and curiosity and, at worst, my lonely reading of what I think to be what you value. You cannot yet easily point me from your page to another site and lead me back to a second page of yours. There is as yet no way to shape a reading of successive spaces, whether visually, socially, or through sequenced links to and from exterior sites. It is a rhetoric of an empty room, the eye listening for a voice, the ear seeking a shadow.

At this point in most discussions of the web someone (on my

campus it is often me) rushes to say that all this is, of course, changing. Throughout the world web researchers and developers are busy developing social and sociable interfaces, Java applets and other schemes that will allow us to map, share, and shape journeys through webspace. Web conferences buzz with technical papers from information warriors with contending models for changing everything, while high above the battlefield public relations barrages from Microsoft and Netscape and America Online burst into clouds of bankable smoke that float like huge balloons casting shadows on everything below. Sometimes the new actually falls like rain.

What we are used to we too often become used by. The web, like most technologies, encourages a constant hunger for newness without a taste for detail. The eye gets tired of watching passing patterns and we settle into a commercial glaze. In that way it reminds me of watching cars as a kid in the 1950s when we prided ourselves on our eye for model years, telling a '54 Chevy from a '53, a Hudson from a Studebaker, with an eye for the broadly iconic that probably owed more than we could imagine to a generation of fathers trained to recognize danger or salvation in the silhouette of an airplane, a Nazi helmet, or Betty Grable. It is the same eye I use now to hit the stop button at the point where the ad is about to load on the *Hotwired* screen or as someone's over-giffed homepage is about to hatch its ostrich eggs.

We are so used to thinking something new will come, and so tired of seeing only patterns, that we never really see or settle into the particularity of where we are. We are discouraged from valuing our own culture and history and representing the shape of our desires for ourselves. This is ironic in a medium whose very stuff is represented history and where everywhere there are remnants of what we once desired to know. Netscape litters our hard drives with the dutifully coded titles of the pages in its cache; we paper our own interfaces with the hotlist or bookmark and strew our own hard drives with the curiously named information farms created by the Eastgate Systems web squirrel software agent. Search engines list page on page of hits, mental footsteps we sometimes mistake for our minds. Some evidences we collect and save, in a process that mimics memory, like snapshots of where we once walked. Others dissolve in the formless track of a Go list, like a walk through a snowstorm, in a process that mimics consciousness.

Metasites (the best-of's, worst-of's, top 100s, worst 500s, Alta Vista, Lycos, Yahoo, Excite, and so on) have moved from utility to commodity and have in fact become the medium. Everywhere there's a cry for better filters as if the point of making coffee were keeping out the grounds rather than whether it tastes any good. People read links rather than sites. At first this development seems alien, it is as if *Readers Digest* suddenly became *Readers Table of Contents*. In fact, however, there is nothing particularly new in this. What's new is how clearly we can see something on the infoscape that before now was either invisible or (what is the same thing) something we became used to.

Almost invisibly in the past, for instance, most library patrons read much more of the online or card catalog entries, book spines, and tables of contents than they read from the volumes themselves. People have only so much time. They can't read everything and so they depend upon others to link them to what they need or wish to read. The notion that editors are "necessary" to filter out the mass of information, of course, insists upon a hierarchy of information and implies one of human beings; it suggests an immanence of cultural values rather than a culture that is constructed by human presence, discussion, and community. As metasites become the medium, neutral tools and interface aspects are not merely programming in the sense of computer instructions but the same programming that spews from the box of televised light on the wall.

And yet the proliferation of electronic discourse can refresh our cultural sense of the value added by human community and may point to a future where the net (in whatever form) confirms our sense of community. For as editorial filters, search engines, and software agents themselves proliferate, we will more than ever need a community (which is to say an embodied presence among others) in which to discuss and agree upon the usefulness of these tools. It has been argued that it was a web of coffeehouse culture and public spaces that led to the emergence of "readable" editorial sensibilities and publishing houses in the eighteenth century. Whether a newly resocialized web of virtual public space can invert cultural evolution and reverse the degeneration of editorial sensibility into tradestyle or publishing house into branded conglomerate remains, quite literally, to be seen.

In so saying I want to begin to make the turn away from the litany of my own unease and toward ways that we can begin to see ourselves in where we are. As I make this turn, however, I want to be clear that I have not Svenned myself, I'm not doing a half-Birkerts from the high board, not suggesting that we Stoll the car and go back to the future, when books was hypertext and Christopher Lloyd drove the car and the quaint band of Swiss physicists built their first hyperlinks in numbered items that riddled a screen of text like holes in cheese. You can't stop the music, thank God, and even when you think you can the kids slip outside into the Delorean and sail away to the land of Studebakers. We have lived through the failure of so-called critical thinking and the media courses that meant to inoculate against buzzwords but that failed with all but the words *media, buzzwords, critical,* and *thinking*. The truth is everything is abuzz:

Neil Postman wanders Newton Minow's desert looking for the oasis while, between issues devoted to giving Nicolas Negroponte more episodes than Nicholas Nickleby, *Wired* interviews a fictional Marshall McLuhan. Meanwhile some of us delude ourselves into thinking that, because Louie's Pizza Palace-dot-Com can put up a homepage with full-color anchovies on a fifteen-dollar-a-month site offered by his local internet service provider, this suddenly puts Louie on the same footing as McDonalds, whose bandwidth is wider than the Red Sea and whose daily McMuffin cash flow alone would wash Louie's feet from under him in a tsunami of melted cheese. Louie, like all of us, is caught between a Hard Rock and *Melrose Place*. If we want to begin to see ourselves in where we are, we will have to find ways to say exactly what we are seeing.

To begin with where we are is in multimedia and what we are seeing is television, a profoundly simple point that Stuart Moulthrop makes so brilliantly that it baffled at least one audience we shared. There is hypertext (which includes hypermedia) and there is multimedia, Stuart suggests. The former is what we do as teachers, writers, artists, learners, while the latter is a variety of television.

Despite my already averred fondness for Marshall and Furuta's cautions to the hypertext community, it is arguable that visually the web set hypertext back at least ten years. The Netscape <frame> structure recapitulates the typical HyperCard interface of 1986 that itself was a version of the typical television interface of the preceding forty years, that is, a box of image (often blond wood) surrounded by tuning buttons. The Hyper-Card interface in turn almost immediately became a multimedia design cliché in CD-ROM where each screen featured a box and buttons (often very like a TV) inside the putty box of the monitor, a scheme that, soon thereafter, itself reinfected television as well so that one now faces the prospect of seeing a box with buttons inside a box with buttons inside a box with buttons. The whole thing becomes what the hypermedia video artist Grahame Weinbren calls "the pit of so-called 'multi-media,' with its scenes of unpleasant 'buttons,' 'hot spots,' and 'menus,' [which] leaves no room for the possibility of a loss of self, of desire in relation to the unfolding drama" (1995, n.p.).

Netscape's frames promise central, supposedly "open" and "changeable" spaces surrounded with immutable (or at least filtering) interface structures that define and mediate the changing experience. You can change the channels but not the commercials. Marshall McLuhan is famous for saying that the subject of every new medium is the medium it displaces, a point Jay Bolter and Richard Grusin (1998) explore in their notion of remediation, in which new media both encompass and seek to make previous media invisible.

Indeed new media are in the business of displacing themselves—not hour by hour in the TV way but cycle by cycle, thirty times a sec, in the way a television screen rewrites itself—and so their subject (happily for them, it is what they know best) is themselves. If what we are viewing is, as some commentators suggest, the marriage of television and the computer, a web is an apt place for it. Like the spiders whose copulation a *New York Times* science section gleefully reported, they are willing to prolong their pleasure by allowing their mates to devour them in flagrante delicto ad mortem con pesto.

Every space is for sale. Cable networks brand the bottom corner of their images with see-through logos while sports network screens carry textual infomercials along their edges as simultaneous contests compete for the viewers attention within the split frame. CNN Headline News does the news in review in the form of a little on-screen shadow play in which an unseen mouse summons story videos from a drop-down menu via a disembodied on-screen arrow; MTV lets geeked boys play Beavis for a byte, by broadcasting music viddies on a split screen over the real-time scrolling transcript of adolescent wit and uppercase cool from their online chatroom.

What's going on here is the familiar encrustation of images that accompanies any holy war, whether mandala, marriage of heaven and hell, World Wrestling Federation, religious icon, American Gladiators, or the matching bibs and banners of the medieval Crusaders. Umberto Eco gained a great deal of media play and mail-list reposting from a piece of journalism (1994) differentiating the Mac and IBM as Roman Catholic and Protestant respectively, thus building a pleasant little religious farce of the same stuff as Marcia Halio's farce of some years ago. Eco was right about the wrong platforms. The current war is between a ragtag group of quasi-Calvinist sentence-diagrammers, strict constructionists, logical positivists, taxonomists, mark-up militia, and bibliographic fundamentalists, the Taggers, versus their Romish antagonists the Flaggers.

Taggers believe that revelation came down from heaven with a TEI header and an SGML DTD and favor baptism through total immersion in platform independence. The Flaggers, an extravagantly baroque group of admen, anchormen, cereal salesmen, and scanner-laden *Wired*-boys, spend most of the day blurring away their blemishes in Photoshop. They believe in transubstantiation, insisting that the image of themselves is in the center of what we see. Because the Flaggers tend to think that literature goes back as far as Bruce Sterling, and that history started with *Star Wars* (the movie, not the delusion of Ronald Reagan), they have ceded control of literature, history, and in fact all archives of everything that doesn't matter (i.e., what's out of copyright or Bill the Gateskeeper doesn't want on his castle wall) to the Taggers and to Sir Henry Chadwyck, the proprietor of all English poetry before Princess Di's phonemail.

The web's become a zombie newsstand filled with news from the front where lots of glossy things wave in the light like Victoria's not-so-secret. It may seem in the interim that the homespun sense of what any of my first-year students know about hypertext—that a new way of thinking and communicating emerges— is lost in the glitter of the war bounty and web booty. It may seem that at this newsstand we don't stand a chance.

Yet because the web is by nature (since it must be writable in order to be readable) partly constructive, and since anyone experiencing hypertext constructs the contour of her own text regardless of the caret-bracketed <frames> that tag and hold her or the trademarked Adobe Acrobats that tumble before her, the possibilities of hypertext will eventually (even in Mr. Bill's market economy) out. It is not simply that, like Louie's Pizza-dot-com, we get to pin up our own shiny fishy things on the newsstand. Rather that for Bill the Bettman to sell us his archive of our images in one-mil increments, or for *Hotwired* to sell demographic nine-digit chunks of zipcoded human genome to Saturn and Sun, or for Netscape to send its paying customers our address in a Magic Cookie, we have to be able to write back, if only in some perversely inverted version of the letters from camp or college, that is, "Dear Dad having fun, wish I were here, I'll send money, signed Someone."

The truth is that we stand a chance. New technologies foster highly articulated horizontal target markets even as they attempt to target them and homogenize them. This suggests that a consuming culture can only sustain itself by eventually providing access across a wide social and economic spectrum. This sort of thinking informs what *Wired* so breathlessly terms Esther Dyson's radical rethinking of intellectual property as a trade-off of content for market segment; it likewise inevitably informed the so-called utopian vision of the former United States secretary of housing and community development Henry Cisneros, who suggested turning public housing into information campuses, and it surely drives the info-industry rush to develop the Yugo computer, the dollar-an-interactive-channel five-hundred-dollar cable modem souped-up TV for home-bound shoppers and popcorn aficionados. Ironically by asserting themselves as a populist/utopian community, web users become an identifiable target market, that is, an audience with definable (populist/utopian) attributes and a hunger for continued access.

To win them you merely have to trade an upstream soda straw for a downstream sewer pipe. It's a pissing contest in which info-marketeers are sure they can prevail. And should the info-communards be flooded from their quaint sewerside homes, they can easily parlay soggy populism into lovely parting gifts, for instance a lifestyle and a T-shirt, or a search engine and a channel guide. Even so it's a game of give and there is space enough to give ourselves to, space enough to stand a chance.

We stand a chance if we create a pedagogy and foster writing geared to making something more than page layouts as game boards with links, something more than house-of-mirrors screen shots of putty-colored boxes and animated buttons. There seem to be two general approaches to doing so. The first (and one I am guiltiest of) is a consciously experimental, stir-fry pedagogy. We treat the classroom as a vast and virtual wok and throw in a mix of web pages, MOO and WOO (web + MOO) sessions, hypermail, interchanges, IRCs, Storyspaces, mail-lists, and a pinch of Quicktime, then serve (sometimes, as I do, all at once) with a cup of steaming Java. Sometimes each ingredient keeps its particular flavor, texture, and color; sometimes it is subgum.

There is an inherent representational problem in an experimental approach since it suggests that someone approaches writing (or any task) with a sense of being experimental, traditional, or what-have-you, doing so in the way that you choose T-shirts or the radio format you listen to. While it may be that some people *do* intend to write in a particular fashion, this intention is political and not artistic or intellectual, and only history can decide whether the effort was experimental, radical, traditional, et cetera.

It was something of this realization that led my Vassar student Josh Lechner to inquire of the author and critic Jane Yellowlees Douglas after a talk she gave on hyperfiction, "If, as everyone seems to want to suggest, hypertext approximates the way our minds work now, would it be fair to call it the new realism?"

Even so, there are advantages to a stir-fry approach to teaching, especially one that insists that synchronous and asynchronous, written and oral, face-to-face and virtual discourses can be integrated with the spatial, analytical, and the visual using hypertext and graphical tools. So stirred, we see writing as spatially as well as temporally represented while remaining aware that closure and coherence are only local and contingent representations. We begin to see ourselves in where we are. We experience writing as what the poet Charles Olson called "field composition."

The second general approach to teaching electronic writing is a consciously theoretical one, a sort of rotisserie pedagogy, where writing is put on the spit and turned slowly over a theoretical fire until, one hopes, the fat drains off while the muscle remains moist. The problem with this approach is that to get the fire you have to burn up the furniture. For the distinction between writing and theory is truly only a matter of furniture arrangement, a convention of academic institutions, course descriptions, and chapter headings. Most writers write conscious of the writing at hand; and theory, which is only another name for this consciousness of what is at hand, has no greater or lesser place than it does in any other kind of writing. In some sense there is never theory, only writing.

A theoretical approach, of course, also has its advantages. It is sort of like consciously thinking about your tongue, at a certain point you either choke or laugh or shift to thinking of your toes. To the extent that hypertext blurs artificial, institutional boundaries, it enables a writing that, even in a theoretical approach, chokes and laughs and wiggles its ears, one that from moment to moment is more and less consciously theoretical, whimsical, practical, lyrical, parodical, and what-have-you, that is, one in which these terms oscillate as what Kate Hayles calls "flickering signifiers" (1993, 71).

I want to suggest another approach, not the inevitable dialectical third, but a middle way, something that plays in the space between and among us, a self-sufficient sensibility given over to others, a lasting presence of particularities that persists beyond the newness we are increasingly used to and too often become used by. To do so I propose that we appropriate as a model for our writing—including and especially those special cases of writing that interaction and collaboration represent—an obscure and foreign sense, the middle voice of the classical Greek verb. The middle voice in Greek is neither active nor passive, yet it offers us a way to see ourselves in where we are. In the middle voice the subject performs the action, but the action somehow returns to the subject, that is, the subject somehow has some special interest in the action. In Greek the middle voice can turn the meaning of the verb "take" *(airo)* to "choose" *(airoumai),* turn the meaning of "have" to "keep close." In Greek the verb "perceive" *(aisthanomai)* is always in the middle voice, always something one does consciously.

The middle voice is a solitary but not a lonely pursuit, what Hélène Cixous calls "walking through the self toward the dark" (1993, 64). In interaction the middle voice returns person to the self, the reader whose experience itself creates the meaning of a new form. In collaboration the middle voice maintains the continuity of the self as something with a voice and name: the writer, a self-sufficient sensibility given over to others. Simultaneously making the action serve the self and the self the action, the middle voice is a voice of coherence, offering us a lasting presence of particularities as a strategy against the fragmentary plenitude and multiplicity that faces and effaces us. The middle voice is a voice given over to coherence: swimming rather than surfing, or, if surfing, body surfing. It has seemed to me for some time that reviving a discussion of coherence (as does Peg Syverson) might add to our understanding of electronic discourse. If we begin with the thesis that coherence can be seen as partially meaningful patterns emerging across a surface of multiply potentiated meanings, then coherence distinguishes itself against other possible coherences, in meaningfulness not meaninglessness. Better still, coherence no longer, if it ever did, distinguishes itself against but rather within. Not against chaos and the random but in recurrence and the flickering.

More recently as I came to collaborate in thinking and writing about coherence with Peg Syverson, Carolyn Guyer, and Marjorie Leusebrink (Syverson et al. 1997), it seemed important to consider the vernacular sense of coherence. In everyday usage coherence remains relatively undetermined and yet admits to inner and outer representations and recognitions (which may be the same thing in our time). In this sense it offers an intuitive term for the distinctive experience most people report in coming to electronic forms as they feel themselves possessed of a different relationship to text and image and a different understanding of how text and image mean.

Coherence has an everyday vernacular sense whose formal qualities are appropriately multiple and morphogenetic (form making). We say of someone, "He doesn't seem coherent" and in some sense mean this to refer to an unspoken and yet recognizable notion of coherence. There is no inclination (as with logic or argument) to test it against axioms, relations, or systematic measures. Yet we also use the word easily and self-referentially. "This coheres for me" is a statement that the dictionary *(American Heritage Dictionary)* definition of "a mass that resists separation" captures well. Coherence in this vernacular sense is very close to what I understand catastrophe theorists to mean by a singularity or phase change: A recognizable shift in which something amorphous takes on form defined by its own resistance to becoming anything other than its own new form. Coherence is the middle voice of consciously making sense for oneself and yet among others.

Yet this is an area that in my experience troubles many of us teachers. We long for a shared coherence, for formats and practices (not to say standards) held in common and for common competencies (not to say requirements) that might inform reading and writing in these new forms. While I am inclined to any conversation about what is common among us, I am less inclined to think that even improvisational commonalties will or should emerge in an electronic age. Here, I think, there is an actual newness upon us. In the place of competencies and commonalties electronic spaces offer the swiftly shifting and easily shared particularities of the middle voice. Instead I would suggest a rethinking of coherence in terms of the successive attendings to persisting forms.

During our English department hypertext seminar my colleague Donna Heiland asks me how I help my students make their hypertexts more coherent. "It is so seductive," she says, "to write these lyric fragments and link them like music. Some of the most interesting hypertexts have a sort of senseless but shapeful beauty and play. I worry that my students will lose their ability to read closely or to argue or to theorize, or at least that they won't be as willing to."

I am midway in an almost unconscious discussion of how we talk about and model hypertext forms in my class (projecting the hypertexts on the screen, for instance, and discussing the way the Storyspaces cluster, then looking at series of links to see if they provide a model of some sort, etc.) when Donna and I each realize (the realization comes in our eyes, a foolishness dawning over each of our faces) that such a discussion may merely be marking the limits of our own literacy. I stop talking in midsentence and we say as much (I literally do not recall which of us said this): "It may be that we cannot see the truly new forms of rhetoric and theory that are emerging. What we see as senseless beauty may be the emergence of as yet unrecognizable new ways of making sense."

As if in confirmation, one afternoon almost a year to the day after Donna and I talked I had three successive meetings with students writing hypertextual senior theses. The first, Shaz, is writing on hypertext and French feminist theory and she has created a four-pole structure, a cluster of thematic tensions (the rational, the natural, arrogance, and boundlessness). Shaz is worried lest each of the sections seem to be too coherent, she doesn't want to be bound by sense, doesn't want the reader to believe that she has escaped the tensions that she has discovered and established as the contrary movements in her text.

Heather, the second student, was writing on contour and consciousness, the nature of authority in texts, and the construction of reader communities and their relationship to feminist discourse. Her writing focused on Nabokov's *Pale Fire* and Shelley Jackson's hyperfiction *Patchwork Girl*. She wanted to learn how to do random links so that the text alternates beginnings with each reading and she wanted to show me how she had established zones of color in the Storyspace where the thinking of certain theorists (Cixous, Retallack, Deleuze and Guattari, etc.) prevails though it may or may not be directly quoted or attributed.

The last of the three, Noah, was writing an extraordinary hypertext narrative of gender construction, a story of a boy growing up in a world of women and a place of presence and multiplicity, of difference and desire. He explicitly rejects the fiction of guard fields and shaped contours exemplified in the Eastgate fictions (and especially *afternoon*) in favor of a fiction of "slippages," "narrative traction," and "release." He wants readers to be free to choose and is experimenting with ways to confront them with this freedom at each turn. He is never unaware of the paradoxes.

"There will come a point in the narrative," he writes, "when the page return will no longer yield as the default reading often encourages a reluctance to engage with the form, the skittish bob of the RETURN [sic] key. This turn constitutes the first narrative 'release,' after which the narrative explodes, or dissolves, depending upon how one looks at it. I've intentionally disarmed the 'default reading' of it's dissuasive availability. In my opinion the page return ought not to encourage the decision not to choose."

I have written elsewhere, in response to those who claim that the so-called MTV generation has no attention span, that in an age like ours that privileges polyvocality, multiplicity, and constellated knowledge a sustained attention span may be less useful than successive attendings. Already in 1934 the poet Ezra Pound felt what we might call the hypermediated urge to "charge language with meaning to the utmost possible degree" (34), including sound and vision, as well as what some have lately and stupidly taken to calling content. Pound's notion of phanopoeisis, or the play between the image as written and read, attends to the oscillation between representation and world (Hayles's "flickering signifiers") as well as the process of confusion and exhilaration as we shift between them. The cave myth in Plato's Republic is as much concerned with what Socrates calls "an art of bringing about [vision]" as it is to the painful and ultimately impossible task of turning from the shadows to the "dazzle and glitter of the light." Hyperfiction embodies this art learned in turning, an art of the interstices.

We experience hypertext fictions as wayward, embodied, and illegitimate. This is the cyborg consciousness, what Donna Haraway calls "an argument for pleasure in the confusion of boundaries and for responsibility in their construction" (1991, 177). Hypertext, even on the web, both embodies and is itself solely embodied by what in print is an invisible process of nonetheless constant waywardness. The reader of a hypertext not only chooses the way she reads but her choices in fact become what it is. The text continually rewrites itself and becomes the constructive hypertext: "a version of what it is becoming, a structure for what does not yet exist," as I put it in chapter 1.

As more and more becomes linked it is arguable that what is passed over becomes more strongly linked on that account. For instance because the web links edgewise, it suggests that every screen is linked to another; hypertext links there thus become the severing of one screen from another (much as each time we read our eyes clip words in sequence). Exclusion and inclusion interact, the outside defines the center; and so increasingly it is not the substance of what we say but its expression and construction that communicates.

It seems to me that these three young Vassar College writers each had a distinctive sense of the shared particularities that construct and communicate vernacular coherences. They were creating texts that require and welcome successive attendings and whose coherence distinguishes itself within other possible coherences, in meaningfulness rather than against meaninglessness. They each insisted on undoing the ways they were used to reading hypertexts. They each expected a reader reading for herself, a reader less concerned with knowing than knowing about.

Teaching and writing in a middle voice calls us to the realization that we have always been less geared to knowing than knowing about. The poet Robert Duncan used to have his students complete a survey that included the question: Name ten master-pieces of literature that you haven't read and know you will never read that nonetheless influence your life as a poet. Years ago I heard someone ask Jean-Luc Godard about the source of the allusive richness of literary reference in his films. He claimed that he absorbed these things while working at a Paris bookstore in the instant between taking a patron's book to the cash register and putting it in its sack. These are descriptions of matrices of meaning. Whether the web or what comes after it will become the poet's bookshelf or the filmmaker's Paris bookstore remains quite literally to be seen. We are in the midst of making a new culture, one we will make as much out of persistence as of newnesses.

Yet we cannot let persistence become habit lest we become used by what we are used to. Nor can we let the allure of newness blind us to seeing ourselves in where we are. We need to bring to our tasks a new and persistent voice, one neither active nor passive, but rather the middle voice that my student Heather Malin invokes in one of the random beginnings that alternate in her hypertext of contour and consciousness:

> To begin and end with consciousness, to ask questions that will not yield answers. This is the way to enter into the movement, to allow room to stretch and become blind, to hear and feel the things that our common sense denies. This is how I come to know the bodies and waves, how I get to the place where my language is really mine, and where I do not worry that I will not be understood. The closure is not the issue, but the (not) getting there. And this is how it begins. (1999)

Chapter 4

The Lingering Errantness of Place, or, Library as Library

Life in the Intermezzo

I come to you as one of you. So, filled with hope, I came to address a national meeting of librarians. We all hope to be one thing or another especially when in strange company; however as someone who was simultaneously a professor of English and the Library (though not a librarian) as well as a hypertext novelist and theorist, the question of whether I came to the library as a wolf in sheep's clothing or a lion lying (in whatever sense one pleases to understand that term) among lambs was not clear at the time either to me or to them. Perhaps I was merely another sheepish Odysseus done up as *outis* (i.e., no man) momentarily escaping an electronic Cyclops.

In my binocular state I knew I shared with librarians a vision that rather than that of *outis* was that of *nomos:* not no man but nomad. We are all of us nomadic creatures of an enfolded sort, whose lives, as Deleuze and Guattari suggest, are conducted in the intermezzo. In my own life I then wore down an uncertain and increasingly virtual path between the classroom and the library, a path like that of Deleuze and Guattari's nomad, "always between two points, but [in which] the in-between has taken on all the consistency and enjoys both an autonomy and a direction of its own" (1983b, 383).

Which is a gentle way to say that we were not yet certain then what a professor of the library does (which was not surprising since after a couple of centuries we still aren't terribly certain what a professor of English does either). Soon after Vassar's then library director (currently university librarian and vice provost at Rice University), the visionary Chuck Henry, announced my appointment we took the title for a test-drive upon a visiting delegation of librarians, faculty, and administrators from another college. The test-drive left no visible tire tracks, the visitors all agreed that a professor of the library was a wonderful thing to have. But this was like zoo visitors seeing an emu: they didn't have to feed me, and anyway, we reasoned, since they were looking to us for answers, what did they know? Not long after, however, the inestimable Dan Atkins, dean of the University of Michigan School of Information, invited me to talk to the students and faculty in conjunction with the publication of my previous book by the University of Michigan Press. Because Dan Atkins is truly a driving force in the library profession, it was my turn to fear tire tracks.

But I was bowled over rather than run over. Dan and his colleagues not only welcomed a dialogue about the uncertain path between classroom and library, but in their own teaching they engaged others in it. In fact over beers following my formal talk I was engaged and fiercely challenged by a wonderful group of School of Information students including two, Nancy Lin and Suze Schweitzer, who for awhile afterward became my active colleagues and collaborators. When I was asked to join other pioneer hypertext thinkers such as Jay Bolter, George Landow, and Ted Nelson in contributing to a "meta world wide web site: a domain devoted to domain design and . . . the larger epistemological concerns posed by . . . hypertext technologies" (Mola Group 1995, n.p.), I thought such a brave new world ought to have many more women than the none it had and I turned to Nancy and Suze as well as my collaborator and partner, the hyperfiction writer Carolyn Guyer. Together with Nancy and Suze's fellow School of Information student Nigel Kerr, we engaged over the network, in MOOspace and on the web, collaborating to create what we described as "a densely linked web of surfaces inspired by (literally: breathing in) the eddying of multiple conversations which have for centuries accompanied quilting and other traditional forms of embodied collaborative art" (Mola Group 1995, n.p.).

Breath and Weave:
On Behalf of Error and Wander

Breath and weave is as close as I can come now to answering the question of what a professor of the library does. Sometimes I knew I didn't just wear down the path between the classroom and the library, but also wore on the patience of my exasperated library colleagues. "What *do* you do?" they asked, and I tried to tell them and in the process we slowly come to wear down what separates us. In this rhythm of ask and answer we learn who we are as an institution and how we sustain the life of the mind among us.

I had been asked to talk to librarians about "The Life of the Mind in the Electronic Age," a title that for many of us I'm sure contains a caesura, a gap, between the mind and the age. Dictionary definitions of caesura suggest that the term may be congenial to the questions at hand. A caesura we are told is "a pause in a line of verse dictated by sense or natural speech rhythm rather than by metrics" *(American Heritage Dictionary)*. Surely any of us would welcome any sensible and natural pause that let us take stock of ourselves in the face of metrics that in an electronic age are most often expressed in dubious or at least frightening orders of magnitude ("a hundred thousand new web pages appear on the network every week"), metrics that do not seem to measure the effect of their additions on the fragile network of our nerve endings.

To be sure the pause of this caesura does mark a gap. Some of us fear that it is an encapsulating gap, the life of the mind trapped in the electronic age like a bee in amber, all its fecund buzzing lost in a static yellow instant of frozen motion. At another extreme some of us hear a predicating gap: a new sense of mind sparks with a hortatory cupola, the gap fills with millennial isness, the life of the mind *is* in the electronic age.

For most of us, however, the gap is aporetic, a gap of doubt, perplexity, multivalency, and loss that seems characteristic of our age: not just the feeling of being off the path (its Greek root *a-poros*) but rather the growing certainty that there might no longer be a path or, worse, that the paths are so multiple that we cannot choose which way to go. The life of our minds seems if not lost then loosed into an aporetic multiplicity. We see shadows that sometimes make us fear that the caesura itself is displaced and that it is we who are lost in the gap.

It little consoles us that being off the path has a noble history, especially in this country, where Thoreau (in a complaint I share) complained at Walden of the "ridiculous demand which England and America make, that you shall speak so that they can understand you," and confessed instead a preference for life off the path. "I fear chiefly lest my expression may not be extra-vagant enough," he said, "may not wander far enough beyond the narrow limits of my daily experience, so as to be adequate to the truth of which I have been convinced" (1997, n.p.). Yet a frequency count of the online *Walden* shows that the words *simplicity* and *simplify* show up five times more often than the words *error* and *wander* combined.

The Emergence of a New Mind

Even so I wanted to speak to the librarians on behalf of error and wander, even knowing as I did, with the sensitivity of one newly one of them, the pride that their profession takes in avoiding error. For there is a kind of error and loss that marks new times and makes new spaces in the midst of the lapsing gap (or last gasps) of doubt, loss, and multiplicity. In the language of the new sciences (to summon Giordano Bruno's phrase) this kind of gap is surrounded by terms such as *autopoiesis, dissipation, phase transition, punctuated equilibrium, turbulence,* or *self-organization*. These terms are the language of chaos, catastrophe, and complexity (as I write this my mind's ear hears someone in the audience say, "No kidding" and I recall James Thurber's line, "I say it's spinach and I say to hell with it").

A new mind emerges within the suffusive gap of stillness in the midst of swirling, of calm moving slowly within roil, of turbulence turned back on itself and yet moving ever onward in changing change. In this suffusive gap both mind and age, both life and electron, feed each other. The word used for this sort of coevolution in the new sciences is not genesis but morphogenesis; morphogenetic systems destroy forms in the process of creating new ones. In the still gap where the phase shift comes, morphogenetic systems move in concert with what is around them, changing within change, mind reminding and remaking itself.

Though this language has deep currency in our most natural sciences, it still seems unnecessarily alien to us. We tend to think of coevolution in terms of the formative myths of our time, monster movies: the thump in the trunk of the *Invasion of the Body-snatchers* that signals the fizzing pod of an other that claims to be us; or Swamp Thing plodding eyelessly out from the electronic muck, an amorphous new mind oozing green in a green ooze like children's toy slime. Yet the suffusive gap is something we are used to, something much closer to the other dictionary definitions of caesura: "A pause or an interruption, as in conversation" or "a pause or breathing at a point of rhythmic division in a melody" *(AHD)*. Music and talk are full of breath and weave, of loss and error. Loss, like breath, is our recognition of time passing; and error, like weave, is our recognition of the linked nature of successive surfaces.

Four aspects of error and wander bear on the continuity and practice of librarianship and the humanities alike, and thus may suggest the breath and weave of the new mind emerging in the electronic age. Yet, as I begin to discuss these four aspects, I would note that I am not in the business of predicting change. In fact I am not only not in any business at all but I also resent the current fashion that urges us each to claim that we are in a business. Instead like most of us, librarians or humanists or whatever, I live in change, living not a business but a presence. As an artist and teacher and technologist I make change and am changed by what others make. It is from that perspective that I want to address these four concerns: the collectable object or the nature of the library; gritty searches or bibliographic instruction; adolescent stacks or the library as publisher; and embodied spaces or the library as library.

One way I wore upon the patience of my library colleagues was with relentless talk of what I call the collectable object, an intentionally polemical term for something that is more a complex than an object. Early on, talking with Vassar's acquisitions librarian, about electronic resources and their collection, she proposed that we might agree on at least one part of a continuum that she saw extending from disk-based circulating hypertext titles to CD-ROMs to interactive websites. The disks she said represented a clear case: we acquire and catalog them and circulate our archive copy. Users must abide by copyright. When the disk returns we clear it and reload the original version on the archive to circulate again.

What about George Landow's *"In Memoriam" Web*? I asked. It is, of course, what we would recognize as an electronic critical edition of Tennyson's poem but with collaborative and hypertextual elaboration. In his hypertext Landow includes the work of other scholars (including Tony Wohl, a historian on our campus) and his own and others' students. What's more in his prefatory materials Landow urges his readers to augment his hypertext with their own notes and experiences. It happened that an English Department colleague was using Landow's hypertext with her students. What would happen if she took Landow's injunction to heart? What if she invited her students (and Tony Wohl) to add to the hypertext? Our normal circulation cycle would lose these additions. We would lose an opportunity to collect a significant record of the intellectual deposit and learning community of Vassar College.

The acquisitions librarian is a thoughtful and creative intellectual. She had questions for my questions: What if my colleague and her students had a bad term? Or what if one class had a better year than another? Or why not save the interactions of classes elsewhere? (At that point it wasn't yet possible, as it is now, to think about what should happen if the *"In Memoriam" Web* moved from Storyspace to the world wide web and back.) What's important about all these questions, the acquisitions librarian's and mine, is that they have no answers except the successive choices, the errors and losses, of our own human community. And that I would suggest is the value that suffuses them and constitutes the collectable object.

An Actual Newness upon Us: Searching for Shared Particularities

In "Coming to Writing," Hélène Cixous says, "I didn't seek, I was the search" (1991, 41). We could say that in the electronic age we don't collect, we are the collection. The value of what we collect is not as much embodied in what it is as in how we found it and why we keep it. If we mistake in what we collect, or if we lack or lose something we should have, our mistaking tells us something of who we are and who others expect us to be. What we do not provide and why forms the permeable, situated boundary of the institution and the constitutive margin of its locality. It is not because we cannot be everything that we choose to do what we do but because we are called to be some thing.

What we felt ourselves called to at Vassar, especially in an electronic age, is a profession of the value of human multiplicity, proximity, and community. As an institution founded on change, our library is an expression of, however oxymoronic it may seem, a tradition of change. The value added by human community is in its being there, and the force of our being (both predicate and nominative) within it. We can tell you where we are and thus assist you in your process of seeing where you are.

This brings me to the question of searches. The value added by human community is in its successive and erring answers to the questions: What do we do with the self? What do I do with myself? which the poet Charles Olson compounds in the question, "How to use yourself and on what?" As the bound blurs between reader and author (in which the merging of library and publisher forms a special case), we feel ourselves increasingly unbounded. The tradition aspect of librarianship called "bibliographic instruction" increasingly takes on the aspects of philosophic instruction. We can only tell you where we are, but no longer can be certain where you are nor say where you should be.

This is an area that in my experience most troubles librarians. Any meeting of librarians rings with fervent calls for common formats (not to say standards) and shared listings of electronic resources. Among my own library colleagues I heard the call for common competencies (not to say requirements) for both student and faculty researchers. While I am inclined to any conversation about what is common among us, I'm less inclined to think that even improvisational commonalities will or should emerge in an electronic age. Here, I think, there is an actual newness upon us.

In the place of competencies and commonalities electronic spaces offer swiftly shifting and easily shared particularities. Web spiders, search engines, and software agents are the cyborgian protozoa in an evolutionary scheme that will take us, humans and machines, toward a coevolutionary world of likewise evolving questions. This evolutionary cell-splitting already increasingly takes the form of what I call gritty searches. More cautious creatures are supplanted by more numerous ones. Smoothly constructed searches are increasingly displaced by successive quick approximations that at each turn are cleansed by iterative query refinements, taking place in virtual and actual communities, involving both computational agents and human beings, and resulting in idiosyncratic and dynamic representations of search and searcher alike. To a contemporary reference librarian such searches are liable to signal a loss of clarity. Even when (or especially because) a machine does most of the floundering, these searches seem wastefully spatial, gestural, fuzzy, haphazard, and physical, and thus gritty in the sense that the particularity of an evolving planet and its creatures are gritty. Yet if they herald a loss it is, I think, the cleansing and morphogenetic loss that engenders a newness.

Beyond Attention Span

There is something lovely about these early days of technologies: the computer is a theater of longings and within it desires that are transparent in print culture suddenly reappear as clearly as the shadows they have always been. The truth is that we have always been less geared to knowing than knowing about. The difference was that the book tallied in physical instances what the network tallies in iterative hits. The errors and losses of suffusing truths are whispered from the unread books upon our (home or library) bookshelves. What I often say (I said it in the last chapter) in response to others who claim that the so-called MTV generation has no attention span is that in an age like ours that privileges polyvocality, multiplicity, and constellated knowledge a sustained attention span may be less useful than successive attendings. Increasingly it is not the substance of what we say but its expression and construction that communicates. The linear, even in the form of traditional information retrieval, is merely a stronger local compulsion. Seen as such, even traditional search structures represent surface-to-surface shifts rather than empirical proofs of the implicit hierarchy of depth. We rest on no single power base but rather learn like dancers, shifting our centers and moving across successive surfaces and textures. We inhabit new forms in the presence and community of others. In a world of shifting centers, meanings are not so much published as placed, continually embodied in human community. This brings me to my next concern: adolescent stacks or the library as publisher.

Barnes and Noble and Other Virtual Realities

In his essay "The Electronic Librarian Is a Verb" Kenneth Arnold links the notion of the "digital university as an information node . . . [in which] organizing information in itself adds value" to "blurring . . . distinctions between library and publisher and author" (1994, n.p.). I gaze into this newly merged blur from the perspective of one who has heretofore mostly been an author and what appears before my eyes is a spanking new pole barn in a strip mall, sided in vinyl clapboard in the faux colonial style of Disney, serving cappuccino and spectacle, and, oh yes, selling books. "I go there first before the library," *Time* magazine quotes a woman speaking about media megastores. Since the stores don't seem to mind if while she's there she reads the books, watches the videos, or listens to the CDs, it is hard to imagine just what they sell her there. More importantly, we might ask ourselves why postcappuccino she would mosey to anyone's decaffeinated library.

A ready answer is that the distinctive extent and character of our collections as well as her own sense of her intellectual community will draw her to us. This answer would be right but, for the moment at least, for the wrong reasons. In a wired world a megastore can extend its collection at will and, one suspects, well beyond our capacities. Likewise character and community is exactly what they sell in a place where you do not have to buy the book to have the dream. Megastores are places of pageantry where patrons play characters in a textual virtual reality (not unlike a MUD or MOO) but in the presence of real objects and among fully rendered neurobiological representations of the other characters (as played by you and me).

In an ideal world any author would want a publisher who could reach both her likely audience and also those readers apt to enjoy an opportunistic discovery of her work. We could call these the certain and the serendipitous readers. Barnes and Noble, Borders, MediaPlay, and so on all try to keep the promises of serendipity and certainty. Their spaces mean to represent the flow from the encyclopedic certainty of the child's library at story hour to the serendipitous adventures of the adolescent's erotic stack at twilight. Within the megastores certainty and serendipity alike are already networked at the level of the distribution channel and the branded multimedia tie-in. What they lack as yet is what the library traditionally has had and what the network promises, the lost community of locality. Still it won't be long before megastores stock virtual localities.

In fact megastores already approximate lost locality and missing self by sponsoring pageants that are not unlike virtual reality scenarios. These pageants take on the guise of television, whether in the form of Star Trek bake sales served up by Spock-eared characters in *Enterprise* garb, or talk shows where quaintly garbed writers in ambiguously gendered Birkenstocks and universal jeans serve up ideas in palatable chunks the size of *biscotti*. Still other pageants masque as self-help seminars where infomercial and food channel celebrities serve up pep talks and nonfat cookies as part of a twelve-step square dance. In these pageants the reader is invited to play parts that range from the child at story hour to the furtive teen who browses *The Story of O*.

In an unbounded time, I suggested earlier, bibliographic instruction increasingly becomes philosophic instruction. As publisher and library merge, philosophy becomes performance, that is, your own sense of which part you would like to play. In lieu of transcendence most of us settle for persistence.

As publishing becomes pageantry and provenance takes the

form of performance, copyright no longer assures and promotes the public's right to access learning but rather provides a stage for the exposition and exploitation of brandedness. In *Wired* magazine Turner Entertainment's so-called wunderkind Scott Sasso says "the creation of good copyrights . . . leverages your content further, higher, faster than anybody else" (quoted in Kline 1995, 111).

What part can we as humanists and librarians play in the midst of this hurly-burly performance art? What comes after the child's certainty and the adolescent's serendipity? What is beyond further, above higher, faster than faster? That's none of our business. I suggested earlier that ours is not a business but a presence. Presence of mind in an electronic age requires persistence. I would like to suggest that the role we might dare to take up as we become publishers of our own pageants is the persistent one of the sacred reader or the adult self. Whether Prospero or Eve, the sacred reader persists in what she reads of the play of self and space, encompassing childhood and adolescence in transcendent performance.

The Embodied Space of the Library as Library

Which brings me to my last concern, the embodied space of the library as library. Everywhere I speak or write I argue the same thing: that the value of our presence as human persons in real place continues as a value *not despite but because of* the ubiquity of virtual spaces. Our embodiment graces actual and virtual space alike with the occasion for value. Patrick Bazin, the director of the Bibliothèque Municipale de Lyon, notes,

> In one respect the "culture of the book"—that is a certain way of production of knowledge, meaning, and sociability—is quite definitely fading a little further from view with each passing day. From another perspective, the syndrome of textuality and its corollary, reading, is becoming omnipresent, and the myth of the universal library looks ever like a paradigm of knowledge. (Bazin 1996, 153)

Bazin's work in building a mixed public space of electronic and embodied forms at Lyon is predicated upon a belief that the myth of the universal library takes place on the actual stage of the library as library. Yet another way I wore upon the patience of my library colleagues was with a repeated mantra: the physical collection must lead us into the electronic collection and the electronic collection must lead us into the physical. I have always been fond of Danish hypermedia theorist Peter Bogh Anderson's playful suggestion for a computer kiosk designed so that as a user explored the space of the museum the movements of the mouse would activate a follow-spot in the museum space. The spotlight would search the actual space and, whether it was within sight of the kiosk or not, illuminate the object depicted on the screen, setting bells ringing and sirens howling, the light dimming and the howl ceasing only when the visitor moved into the actual space to silence the longing object.

The mind of the electronic age must move out into the world. Arnold reports Todd Kelley's description of Project Muse as an effort "to extend the educational reach of the library beyond the walls of the institution," and he rightly suggests that it represents an instance of "a new kind of librarian" (1994, n.p.). Yet it is important to recognize that extending the walls outward likewise takes the world in. The new librarian, the sacred reader, takes the world into a real place that is neither a mythic universal library nor, for that matter, merely a digital one. Hypertext theorist, systems programmer, and fiction writer Cathy Marshall and her colleague David Levy write in the digital-libraries issue of the *Communications of the ACM:*

> The academic and public libraries most of us have grown up with are the products of innovation begun approximately 150 years ago. We would find libraries that existed prior to that time largely unrecognizable. It is certain that the introduction of digital technologies will again transform libraries, possibly beyond recognition by transforming the mix of materials in their collections and the methods by which these materials are maintained and used. But the better word for these evolving institutions is "libraries," not digital libraries, for ultimately what

must be preserved is the heterogeneity of materials and practices. (Levy and Marshall 1995, 77)

Two Pragues, or,
The Lingering Errantness of Place

The heterogeneity of our materials and practices suffuses us in
music and conversation, breath and weave. I want to end with
the music of two voices that move weaving out into the world
and breathe it in. After demonstrating the Internet Public Library
at the world wide web conference in Darmstadt, new librarian
Suze Schweitzer visits Prague and sends this email to her collaborators on the Mola project:

> I am sitting at an old desk in an old office tower which is called
> the Motokov building and used to be the place where the old
> regime took care of all the bureaucracy surrounding the export
> of Skoda cars (skoda in Czech also means "too bad, it's a pity").
> Ken, who works here now at [the Czech website], informs me
> that the surrounding neighborhood (the outskirts away from
> the old city . . .) is an excellent example of "socialist realism"
> that means grey brick and metal buildings in boring repetitive
> patterns that dampen all enthusiasm and apparently also any
> signs of life—no trees squirrels birds children, etc. so in the
> middle of all this grey, and next to a window on the sixth floor
> that overlooks a construction sight that has been a construction
> site for over eight years and probably will be a construction
> sight for quite some time because no work has been done on it
> since Soviet money left, I finally get a relatively speedy internet
> connection, which seems solllllluxurious, and I open telnet and
> Netscape at the same time and . . . see the flurry of messages
> from you all about the project, and I am touched to read that I
> have been consciously included and begin to feel that tingling
> sensation in my nose that comes just before crying . . . partly
> because I could not be there, here, somewhere, to participate;
> partly because it seems I was a participant and I am touched in
> the same way a person is touched when she receives flowers
> because it means that another has been thinking of her when
> she was not around, and also because she can enjoy the flowers.
> (Personal communication, email)

We must move out into the world, reading it simultaneously from above and within. Not far off from where Suze sat, it happened that my own son Eamon sat at the same time, also in Prague, visiting as part of a senior year in high school spent as an exchange student in Dresden. A male Miranda in email he too saw a brave new world within a Prague that was "da bomb!! it just leapfrogged london and rome as my favourite euro-city . . . indeed *wild* but in a good way, just a lot of stuff that one never sees anywhere else. like hospitality and international friendliness for example. everyone young (and bums too) just sits out on these statue stairs in the middle of the city and drinks beer and talks to each other, that goes on until 4am, germans, czechs, swiss, italians, spainish, danes, dutch, swedes, english, irish, scots, lithuanians, and americans."

For me these email messages likewise suggest how in the face of the voracious newness of the web with its nomadic hits and Virillian speed, we might interpose the lingering errantness of place, the heterogeneous practice of culture as the experience of living in a place over time, with each word sounded and suffusing, each a caesura, marked and energized: experience living place time library mind

Chapter 5

Beyond Next before You Once Again: Repossessing and Renewing Electronic Culture

Color

We are who we are. We are used to saying some things go without saying. This does not. For it is the saying that makes us what we are.

This essay borrows as its subtitle the name of Sherman Paul's collection of "essays in the Green American Tradition," *Repossessing and Renewing,* as a conscious nod and a continued memorial to my mentor, who late in his life offered me the grace of affirming that my hypertextual experiment was for him within the Green Tradition. I also appropriate the title as a charge to myself to take up Sherman's journey in the face of an emerging electronic culture seemingly too ready to discard place, body, and history. Notions like net years and virtual presence threaten the persistence of being that the tensional momentum (to use Carolyn Guyer's phrase for the reciprocal aspect of what we otherwise misrepresent as polarities) of repossessing and renewing calls us to. This essay intends a gesture toward what comes beyond next, which is nothing less than what is before us: ourselves as expressed within time and space. We are who we are and we see ourselves in brief light but live always in the shadow of what comes next.

We are surely not the first but without doubt the most self-conscious age to see ourselves as living before the future. In our technologies, our cultures, our entertainments and, increasingly, the way we constitute our communities and families we live in an anticipatory state of constant nextness. There is, of course, a branch of philosophy that concerns those who see themselves as inhabiting the time before the future. That branch, eschatology, is perhaps the archetype of othermindedness and its itch of desire for constant, immediate, and successive links to something beyond.

Eschatological ages have both their virtues and their particular vices. The chief virtue is hope, that constant anticipation of the next that keeps us poised, unsettled, and open to change. The chief vice is paradoxically inaction, a self-satisfied belief that there is no need to act in the face of a decisive and imminent history. Like any teacher and writer, I see my task as encouraging virtues and discouraging vices insofar as I can recognize the difference between them. And so as a teacher and writer deeply involved with technology I have for some time been concerned with the passivity that electronic media encourage.

Early on I distinguished between two kinds of hypertext, the merely exploratory and what I termed the constructive hypertext, seen as "a version of what it is becoming, a structure for what does not yet exist." More recently as both packaged infotainments on CD-ROM and the world wide web alike have encouraged a kind of dazzled dullness and lonely apprehension, I have proposed in chapter 3 that we appropriate as a trope, if not a model, for our interactions an obscure and foreign sense, the middle voice of the classical Greek verb. The middle voice is a form neither active nor passive, yet one that tips the meaning of an action to account for the presence of she who acts or is acted upon.

Our sense of ourselves as actors colors our appreciation of the world in which we act. We are who we are in an active and public sense. We become both the beneficiaries and the constitutive elements of what we might call, to use an old-fashioned term, the public good.

In its eschatological aspect (and perhaps in millennial fervor as well) the web encourages at least an expectation of public goods, if not a public good. There is a widespread if naive expectation that material ought to be universally and freely available. "Content-producers" (the obscene worker-bee appellation for artist and writers and thinkers) are urged by commentators like the computer market analyst and erstwhile pop-philosopher Esther Dyson to find their incentive and make their living from the value added in lectures, sinecures, and so on that result from public knowledge of their work. What makes such urging suspect is not its truth value—since what Dyson and others so breathlessly prognosticate is merely the yawning present state of most artists and intellectuals—but rather its misprision.

Here I mean misprision both in the common sense of that word as something of an insult and the root sense of misprision as a maladministration of public office. For the truth is that the kind of economy that would provide incentive and sustenance to she who provides free value to it assumes a common understanding of the public good that free access to information, knowledge, and art represents.

The question at hand seems to be whether there is any longer a Public in either the civic sense or economic sense. The public's expectation that it will have free access for possession of public good(s), cultural or otherwise, is fundamentally constructive. Art and commerce each intend to serve freedom (or at least make that claim). Yet to the extent the web is predicated on anonymity and irresponsibility, no publics actively assume the responsibility for the goods to which they have access. Instead they passively allow it, in greater and lesser volumes like irrigation sluices. So-called value-added schemes (the inner sanctum, the registered shareware user, and intranet) induce this public to increase the inward flow, to let the supposed provider include knowledge of the public holdings. In the net economy you don't take money from people, you give them the right to let you in the place where they spend it. When you charge access on the net it is the same as doing advertising, just a matter of what people will let into their lives.

As artists and thinkers and teachers we want, I hope, to reverse the flow. We want to encourage responsibility for even seemingly passive choices, for virtual worlds, and for alternate selves. We want to encourage a collaborative responsibility for all that we as makers and shapers consider a desirable thing to maintain and for which, we believe, there exists if not a Public then various communities willing to sustain it.

This is to summon an othermindedness that is less a focus on the other than upon our mindedness. Networked learning calls us to be mindful of ourselves in increasingly other roles than that as passive consumer, but rather as cocreator and reciprocal actor.

Lately I find it useful to ask anyone I speak to, but especially my students, to consider what comes next after the web, not in the sense of the next browser increments, Java applets, and operating system transparency, nor the next order of magnitude of increase in instantaneity or availability. At first it is a shock—especially for those who have not lived through the succession of vinyl to cassette to CD to DVD—to understand that I do not mean some mere appliance like the cable-bound network computer. Instead I mean what next literacy, what next community, what next perception, what next embodiment, what next hope, what next light.

Perhaps these are the old habits of a once Irish Catholic boy, or the new habits of an increasingly old-hat hypertextualist, but they are also habits of othermindedness and, while not restricted to any techne, are characteristic of the way we see ourselves through our technologies. Thus, for instance, the Canadian painter and theorist Guido Molinari turns a color theory into a networked otherness:

> Establishing the capacity of color to bring about an indefinite number of permutations is what, in my view, constitutes the dynamic that produces fictional spaces and gives rise to the experience of spatiality—excluding, by definition, the notion of any specific, given space. It is only through the notion of becoming which is implicit in the act of perception that structure is explored and established as existential experience. (1976, 91)

We are who we are. We see our spaces in how we live our differences and we live in what we see of ourselves within their otherness. This is both the present task and the constant teaching we are called to by any techne from the oldest days to the next days that, after all and despite our lights, can only follow the present as we perceive it.

Body

We seem to have lost track of mortality, if not death, in the face of the constant replacement that is characteristic of electronic text and culture ("print stays itself, electronic text replaces itself"). We know better but we wish for more.

The body is the fundamental instance of a nextness that argues for the value of what has come before it. It grounds and forms the "existential experience" that Molinari characterizes as "the notion of becoming which is implicit in the act of perception." Because we are going to die, we are the embodied value of what has come before us. I mean (you mean) the ambiguity of "come before" here, both the sense of that which—and those who—precede us, and in the sense of what we sense, as in that which comes before our eyes. In this instance, it may be useful to redeem the euphemism. Because we are going to pass away, we are the embodied value of what we pass through and what passes before us.

It is the push of passing, the fixed stamp of ourselves, that we resist in our embodiment. All this passing leaves us open. "Location is about vulnerability and resists the politics of closure," says Donna Haraway; "feminist embodiment resists fixation and is insatiably curious about the webs of differential positioning" (1991, 196).

In this particular eschatological age we cannot help hearing the present state echoed (or prefigured) in Haraway's use of the word *webs*. Yet I would argue that the solipsistic perspective of self-selection that thus far characterizes the brand-name world wide web (so-called in a time when even ketchup bottles have their own URLs) falls short of embodying the curiosity that drives most of us to it. Also, and more importantly, the web fails as yet to render the "differential positioning," the moving perspective (pun intended) from which Haraway can claim, "There is no single feminist standpoint because our maps require too many dimensions." The current web fills the sweet emptiness of space with static and keeps us static in the flow of time.

We are who we are and we stand beside a river. When my Vassar colleague and fellow Sherman Paul protégé Dan Peck told me the news of Sherman's final diagnosis, he urged me to write him but wisely warned me against the elegiac in favor of newsiness and shared thinking. Despite Dan's fraternal concern, it was unnecessary advice in the sense that I could not in any wise take it. In my mind, and given my own quasi-Irish predilections, the only news is our mortality and the nature of all shared thinking is elegiac. We are used to saying some things go without saying, but it is the saying that makes us what we are. "Whoever wants to write," Hélène Cixous suggests,

> must be able to reach this lightening region that takes your breath away, where you instantaneously feel at sea and where the moorings are severed with the already-written, the already-known. This "blow on the head" that Kafka describes is the blow on the head of the deadman/deadwoman we are. And that is the awakening from the dead. (1993, 58)

My tone with Sherman had always been excessive and elegiac from our first encounter in his office where I begged admittance to his Olson-Creeley seminar claiming the survivor's rights of someone who had failed to honor Olson during a Buffalo youth and now felt the blow on the head. My recollection (very clear actually) was that Sherman shared his own story of (literally) overlooking Olson across Harvard yard, thus taking me into the seminar while surely more-deserving, if not necessarily better-suited, graduate students were left outside. Likewise Sherman's tone had always been a survivor's and one of shared perspective, looking outward like the figure of Olson's epic Maximus poems. While Sherman may not have used these terms exactly, he often thought about what Haraway calls "resisting the politics of closure" and "differential positioning." Thus when he came to collect (in *Repossessing and Renewing*) his introductory essay to *Walden*, he meditated upon survival and being, casting the question in terms of how we live open to a world in which we are enclosed by responsibilities and the demands of others:

> Writing itself opens a space truly one's own, and when one enters it he is no longer moved by pressures of survival or ambition, but by the wholly different, imperious pressures of intellect and art. Personally, there was nothing paradoxical about my writings about Thoreau: it allowed me, as the classroom did, to live in my vocation, and gave me a way of being-in-the-

world and the well-being without which the academic situation would have been less tolerable. (1976, 55)

This living-in is what constitutes location on Haraway's, Olson's, or Sherman Paul's terms, and what Haraway means by an "embodiment [that] resists fixation." The paradox, of course, is that such an embodiment is bracketed by the saying that cannot go without saying, the elegiac voice that makes us what we are. "Could it be," Sherman wrote in the same afternote quoted above, "that Life and our lives, the two words that enclose the [collected] introduction to *Walden,* were fortuitous?"

Not often an ironist, Sherman had a mortal ironist's retrospective sense of the tensional momentum of ambiguousness of the word *fortuitous,* with its paired qualities of happenchance and lucky legacy. He knew that the young man who by happenchance began his energetic scholarship with Life in the uppercase abstract had been lucky enough to live to a point (not then the end) where he could see the closure of life as lived and bracketed in ourselves.

It is this same bracketing that my old friend Janet Kauffman means to summon in her novel *The Body in Four Parts.*

Deprived of the elemental world—and who isn't, with a globe divided, the whole planet sectioned, roofed, cut and pasted— even its waters—what can a body do, if it is a body, but acknowledge, salvage, the elements in its own boundaries. Draw them out. Wring them out. House. Host. . . . [Summon] its lost geographies. (1993, 12)

Writing to Sherman at a point that bracketed his mortal life, and thus marked the fortuitousness of my own, I was convinced of Kauffman's claim that "it is the dream of the body—to know a place bodily and to say so" (1993, 119). That is, I was convinced that the important questions facing us as an increasingly technological culture will be played out in places like Vassar and similar human communities where we consider and profess the value added by (and embodied in) that community. In my last letter I tried to tell Sherman how despite (or perhaps on account of) my modest role in its development, it seemed to me that the pervasiveness, immediacy, and unmoored multiplicity of electronic culture will inevitably and increasingly throw us back upon human communities as sources of value, identity, and locality.

By that time we had moved to New Hamburg, one of the few towns along the Hudson where the railroad runs on the right side of town and not between town and river. Thus I was aware, as I also told Sherman in this last letter, that although we were only a block and a half away from the river, we were a lifetime away from understanding even the simplest of its rhythms. I was reminded of how in an almost identical context—discussing Barry Lopez's *River Notes*—Sherman had quoted the poet Charles Bernstein about the archaic and its "chastening lesson . . . of our own ignorance and the value in acknowledging it" (1992, 85).

Sherman wrote me back on Easter morning. The crows, he said, had dusted the snow from the branches of the pines. He was feeling briefly better. "There is no assurance that this well-being isn't transient," he wrote, "but isn't the transient, even miracles, which I am beginning to settle for, in the nature of things?" He had been able to walk out, he said, and "inspect the trees I've planted, some 35 years tall, and observe the emerging spring."

He once wrote me that over the years he had planted fifteen thousand trees throughout the eighty acres at Wolf Lake in Minnesota. I do not think it was an exaggeration. In some sense I am among them.

Wood

The crows dust snow from the pines.

What, finally, are we to make of the fundamentalist aspects of what seems a wood-pulp fetishism among the postlapsarian (I won't call them neo-Luddites, Ned Ludd's fight is my fight as well: we are who we are, we have bodies that the machines cannot deny) critics of new writing technologies? Already of course my rhetoric barely hides its contradictions. Yet to convey and hide its contradictions in the same gesture is, of course, the purpose of any rhetoric, any tree, or, as we shall see, any screen alike.

We are "finally" to make nothing. Or rather we finally make only ourselves. Yet these selves are made of nothing lasting, wood or otherwise. In the face of such knowledge, or perhaps despite it, it seems that these contratechnologists—the postlapsarian and eschatological wood sprites—long not to last but to be among the last. In an age of constant nextness they long to set the limits: write here but no farther, write so that the mark is read in carbon but not in light. In an online exchange about "the cultural consequences of electronic text" (which he contributed to by the faux network of proxy fax) Sven Birkerts seeks to set such end terms:

> I catch suggestions of the death of the natural and the emergence of proxy sensualism, one tied up with our full entry into a plasticized and circuitized order. These synthetic encounters could only become real pleasures—objects of rhapsody—after we had fully taken leave of our senses (literally). . . . A utility cable will be beautiful (and not in the surrealist sense) because we will have lost our purchase on branch and vine and spiderweb. (1995, n.p.)

The prose is felicitous and rings round like a vine, yet the thrust of what he circles becomes clear upon further viewing. This is a maypole ceremony, a self-garlanding. He seems at first to come (literally) out of the woodwork with the claim for fetish. His stance seems to be that the book, being vegetal (i.e., made of wood), assures that we will continue to inhabit a natural world. Yet the obverse claim, that is, that the book in its apparent naturalness has blinded us to vine and spiderweb, is not only equally likely and as easy to sustain but also has been made by both the great men Birkerts admires from Plato to Thoreau and by a woman whom he may and I do admire, Donna Haraway, whose "webs of differential positioning" are considered above.

What really underlies Birkerts's argument, like most reactionary polemics, is I think a profound distrust of the human community and the future. We seem called upon to believe that, because there are apparently no naturally occurring polymers (let us put aside the natural origin of the copper—or the gold!—of the computer's utility wire), Birkerts or my granddaughter will abandon the grape arbor for the world wide web. I take another view. The so-called real pleasures of synthetic encounters are just as likely (in a world in which we trust our progeny) to call them more strongly to the real pleasures of human community and the world around us. To claim that the natural world will necessarily be transformed beyond recognition is proxy sensationalism and impure fetishism. It is just as likely that the natural world will be transformed (which is to say brought back before our eyes) into recognition and that we shall gather there (by the river), not in rhapsodic flight from the net, nor in leave of our senses, but within the leafy garden of forking paths.

Though how we see ourselves as clothed in the natural world (whether shamed into fig leaves or in the splendor of the grass) is an old story and depends upon our understanding of tree and garden alike. In Haraway's explicitly post-Adamic paradise, "Webs can have the property of systematicity, even of centrally structured global systems with deep filaments and tenacious tendrils into time, space and consciousness . . . knowledge tuned to resonance not dichotomy" (1991, 194–95).

The turn from dichotomy to resonance is not easy and requires us to see ourselves proprioceptively, that is, inside out. Regis Debray seemingly makes a more reasoned case for the fiber book as symbolic object rather than a fetish:

> Written text converts the word into surface, time into space; but a single graphic space remains a planar surface. Written text, like screened text, has two dimensions; a parallelepiped has three, like the world itself. The memory of the world, materialized in the book, is itself the world. . . . A volume of paper and cardboard is a resilient and deepening microcosm, in which the reader can move around at great length, without getting lost in its "walls." The book is protected because it is itself protective. . . . One can take one's lodgings there so to speak, even curl up comfortably. (1996, 147)

Yet to a feminist critic this microcosm where the homunculus "can move around at great length, without getting lost in its 'walls' . . . [and] curl up comfortably" must sound (in the root sense) familiar. It is the place where the family is formed, the inside-out that makes us who we are. To paraphrase the title of Irigaray's famous essay, this book which is not one is the multiplicity of the room as womb, not the tome as the world's tomb. The memory of this world, materialized (and maternalized) in ourselves, is itself the world. We are who we are.

Debray's claim (or my appropriation of it here) requires that we read ourselves from without (our lack is that we are one) and thus open ourselves to who we are within (where the difference between who and whom—and womb—here is everything). This requires a sense of not merely our not-oneness but our doublenesses. Doubleness of course also recalls Irigaray's essay, in which "within herself, [woman] is already two—but not divisible into one(s)—that caress each other" (1985, 24). In this doubled sense our memory of the world—and thus of what the book means to enact and the screen aspires toward—is neither an occupying gaze nor a phallocentric taking up of lodgings but rather the to-and-fro flow of meanings in which "the geography of . . . pleasure is much more diversified, more multiple in its differences, more complex, more subtle, than is imagined" (1985, 103).

"While the noun *screen* connotes an outer, visible layer, the verb *to screen* means to hide," the poet Alice Fulton writes in a meditation on the nature of electronic texts:

> The opposing definitions of screen remind me of stellar pairs, binary stars in close proximity to one another, orbiting about a common center of mass. Astronomers have noticed a feature common to all binaries: the closer the two members lie to one another, the more rapidly they swing about in their orbit. So screen oscillates under consideration. (1996, 111)

The place where binary stars lie is, of course, a bed. We are embedded in our differences and we oscillate under consideration. "Genuine books are always like that: the site, the bed, the hope of another book," says Cixous.

> The whole time you were expecting to read the book, you were reading another book. The book in place of the book. What is

the book written while you are preparing to write a book? There is no appointment with writing other than the one we go to wondering what we're doing here and where we're going. Meanwhile, our whole life passes through us and suddenly we're outside. (1993, 100)

In that sudden we read ourselves from without and thus open ourselves to who we are within. What has happened to the wood? the reader might ask. We might misread Shakespeare but not necessarily our natural grain to think that we are as much born into the wood by our mother Sycorax, as born from it by our stepfather Prospero, whose words and books are after all our only evidence that we were trapped there. In any case, whether we are fathered by tempest or a grim fairy tale, our truest nature (or at least our dream) is that we have to move. "In order to go to the School of Dreams," Cixous says,

> something must be displaced, starting with the bed. One has to get going. This is what writing is, starting off. It has to do with activity and passivity. This does not mean one will get there. Writing is not arriving; most of the time it's *not arriving*. (1993, 65)

Not arriving, where have we come to? We can respond affirmatively, even enthusiastically, to Debray's claim that "The technological ecosystem of the textual relates back—in the same way as any microsystem—to the wider scale of cultural ecology," and even accept the proposition that he suggests leads from it as "something that bears a strong semblance to an anthropological constant: human communities *need* a unique defining space to belong and refer to" (1996, 148). However, doing so does not, I think, oblige us to submit entirely to his further, enigmatic claim to "formulate it all too laconically: *no culture without closure* (and time alone as the defining medium of anything) *cannot close it off*" (148).

There is a closure that does not close us off but that, while leaving us open encloses us. "Skin wraps body into a porous and breathing surface through which a variety of exchange takes place," the artist Heidi Tikka suggests (1994, n.p.). Tikka suggests a notion of inter/skin as a correction to the penetrating phallic gaze of interface. Skin, she says,

> covers the face as well, but the communication skin participates in: touch, secretion, receptivity and sensitivity—when blushing, having goose pimples, shedding tears or sweating—remains the underside of human communication. The incalculability of these signs prevents them from being valid currency in the phallic exchange. In the economy of phallic representation skin does not count, it functions as a material support.

Skin is screen. "I think about these things we create—these hypertexts—as part of our skin," Martha Petry argues, "as permeable and open as the eyes on our faces. . . . what we see here . . . is the outer membrane, the surface layer, the rind or peel of fruit, a film on liquid" (1992, 1). Tikka evokes Irigaray explicitly—and both Petry's and Donna Haraway's notions of permeation implicitly—in arguing that "an inter-skin has a great sensitivity and completion for receiving a variety of signals from the environment and capability of changing its state accordingly" (1994, n.p.). This is a literacy that offers us both well-being and the being in the world that Sherman Paul summoned from Thoreau; one that rather than leaving us, in Birkerts's terms, "fully taken leave of our senses (literally)," instead for Tikka sensually "connects with other surfaces and conducts and circulates information in a network of similar surfaces."

In the place of Debray's laconic formulation "no culture without closure" we are faced with a Lacanic counterproposition of encompassed enclosure. Birkerts's fear that we will take leave of our senses is posed as a fear that we will lose sight (of ourselves). Yet it really is a fear that we will lose touch with parts of ourselves. "The contemporary pressure toward dematerialization, understood as an epistemic shift toward pattern/randomness and away from presence/absence," N. Katherine Hayles suggests,

> affects human and textual bodies on two levels at once, as a change in body (the material substrate) and a change in the

message (codes of representation). Information technologies do more than change modes of text production, storage, and dissemination. They fundamentally alter the relationship of the signified to the signifier. Carrying the instabilities implicit in Lacanian floating signifiers one step further, information technologies create what I call flickering signifiers, characterized by their tendency toward unexpected metamorphoses, attenuations, and dispersions. (1993, 76)

The fear of losing the world is a fear of dismemberment; we close ourselves off into the zipped, conservative ground of the male gaze and colonial vista alike. Against such a fear of loss there is the countervailing play of surfaces, the joy of several worlds at once, passing and multiple. The "inherently diffuse surface" of skin, says Tikka, "changes identity, sometimes dissolving itself into another surface in a way that makes the identification between the two impossible . . . [and] refrains from the production of a fixed subjectivity" (1994, n.p.).

In place of the male orgasmic rush of rhapsody there is the fugal female orgasmic of not-arriving; in the place of Birkerts's "purchase on branch and vine and spiderweb" (where "purchase" is a noun of knot and lever and gather), there is the weave (the textus) of unexpected metamorphoses, attenuations, dispersions, and the unmoving silence upon which Ezra Pound ends his *Cantos* (1954):

> I have tried to write Paradise
> Do not move
>   Let the wind speak
>   that is paradise.
>
>   (Canto CXX)

Light

We hear the wind through the trees as whispering music but we read it as varieties of light. In the play of inherently diffuse surfaces we hear the world speak.

Before the book of fiber there was the book of skin, whether the vellum of the codex or the earth's own skin, clay tablets worked in dampness and dried in wind and light. The mediums of exchange for the skin are light, air, and water. Let us examine them in order, or rather as if they had an order.

The wood-pulp fetishism of postlapsarian critics seems at first a mistrust of the eye and a privileging of the hand. Their longing for the "resilient and deepening microcosm" of paper and cardboard seems a wish to touch the wound of culture and in that gesture heal over the openness that is its possibility. Yet there is a sense of reading that seems to favor the eye and mistrust it in the same gesture. In fact it mistrusts gesture, which is after all the work of surface, and thus demands to inscribe it in the mark.

A year ago the wind of descending helicopters spoke through the bare winter trees upon the campus where I teach and thereafter I saw this mistrust in action. My Vassar College colleague Don Foster, in the course of using computer tools to establish Shakespeare's authorship of *A Funeral Elegy,* had drawn international media attention. Now the media had asked him to turn his attention to another, then more notorious, anonymous authorship, that of the political satire *Primary Colors.* In writing about Shakespeare's text, Foster says,

> *A Funeral Elegy* belongs hereafter with Shakespeare's poems and plays, not because there is incontrovertible proof that the man Shakespeare wrote it (there is not) nor even because it is an aesthetically satisfying poem (it is not), but rather because it is formed from textual and linguistic fabric indistinguishable from that of canonical Shakespeare. Substantially strengthened by historical and intertextual evidence, that web is unlikely ever to come unravelled. (1996, 1082)

Yet what served for Shakespeare and brought Foster his scholarly reputation and media notoriety alike did not serve entirely for the author of the political satire. The helicopters had come because Foster all but conclusively identified the author of *Primary Colors* as the *Newsweek* writer Joe Klein, a story that CBS News and *New York* magazine reported in February 1996. Yet it was not until the following July, when the *Washington Post* engaged a handwriting expert to examine handwritten emendations on the galleys of the novel, that Klein and his employer owned up.

We might mark this down as a minor mystery, a passing event in the history of literacies and the further adventures of a premier Shakespeare scholar and technologist, were it not for what it suggests about the postlapsarian insistence upon the place of marks. Foster couches his own methodology in a positivist science in which "researchers can now test . . . matters [such as authorship] objectively, by mapping the recorded language of an archived writer against the linguistic system shared by a community" (1996, 1083). We can put aside for the moment the question of authors whose works are not archived (or indeed whether archives of pulp or of light are likely to be more lasting than Horace's bronze), and we can even defer the question of where and how we find the marks of community, to ask a more fundamental question.

Does the mind leave a mark?

This question is of course another way to address our mortality, the mark we leave upon the world. Is the person in the physical mark or the mind's mark?

Foster's screens played across the body of text and yielded light. The methodology for Shakespeare was the same methodology used for the lesser scribe, locating "an extraordinary match between the distinctive vocabularies [as] a function not principally of verbal richness but of individual preference or habit . . . [as well as] fairly ordinary nouns used as only [the author] is known to have used them" (1996, 1083).

We can of course see neither match nor preference, neither habit nor the idiosyncratic and thus not either the extraordinary ordinary, in a single screen or even any sequence of them. Unlike the characteristic whorls and slants that are the handwriting analysts' stock in trade, the mark of the extraordinary ordinary flits across a screen in instances of light whose recurrences mark Foster's web of "textual and linguistic fabric."

A liar may not own up to a fabric of light (itself another name for skin). Nor, it seems, might a postlapsarian. Both however seem susceptible to certain carbon forms, dried pulp and the etched mark. This mistrust of light on the computer screen is, I would suggest, a variety of our mistrust of the body in and of itself. To the extent that light and its dimming and recurrences mark the temporal, it is likewise a mistrust of our own mortality. Finally, it seems a mistrust of the locus of meaning that, as Foster's methodology suggests, is shared by a community. We cannot be sure what we see except in community. For what we see, as René Angelergues suggests, is itself woven with what we have not been able to see:

> Perception, hallucination, and representation are part of the same process. The object to be perceived is in no sense an "initial condition" that creates a causal chain ensuring the object's imprint (image or information) in a focal centre, but rather a complex and conflictive process that mingles and opposes knowledge and recognition, discovery and familiarity. (Cited in Ottinger 1996, 26)

I believe the mind leaves its mark in the light filtered through the snow-dusted branches of thirty-five-year-old trees. In some sense I am among them.

Air

We are afraid to find ourselves in air. Dreams do this to us, as do leaps, journeys, syntax, the weave of perception, hallucination, and representation, the book the web and the network as well.

Wind is sound. Recurrence is the sounding of memory in air. Air is *spiritus*, breath, whisper, ghost.

We have talked about all this before. We are who we are. We are used to saying some things go without saying. This does not. For it is the saying that makes us what we are. Recurrence is the sounding of memory in air.

This is child's play. Anyone who has read my writings about electronic texts recognizes a characteristic, not to say obsessive, rhetorical stratagem in them (and thus here). The recurrence (sometimes what we call "whole cloth" though we mean patchwork) of a phrase or paragraph (and at various times as much as a page). Self-plagiary is proprioception. Anyone who has read my writings about electronic texts recognizes the recurrence of Horace's phrase from *Ars Poetica* in them (and thus already above here): *exegi monumentum aere perennius.* "I have built a monument more lasting than bronze."

"As children," write Cara Armstrong and Karen Nelson, "we experience space through all our senses and we have an intimacy with place. Through monuments and rituals we try to recall this intimacy and awareness" (1993, par. 1).

The mark of light is sounded in recurrence. That sounding is the body's surface. These short sentences form a pattern not an argument. Its monument is what Lucy Lippard terms *overlay:*

> It is temporal—human time on geological time; contemporary notions of novelty and obsolescence on prehistoric notions of natural growth and cycle. The imposition of human habitation on the landscape is an overlay; fertility—"covering" in animal husbandry terms—is an overlay; so are the rhythms of the body transformed to earth, those of sky to the land or water. (1983, 3–4)

Overlay likewise offers a sense for understanding what, in a discussion of a student's (Ed Dorn's) work in terms of *his* mentor and teacher (Charles Olson), Sherman Paul discovers (he writes this book, *The Lost America of Love,* in short sentences that form a pattern, not an argument) in Olson's sense of

> *Quantity.* Olson says it used to be called *environment* or *society.* He doesn't elucidate. Perhaps he suggests enough when he says it's the present time, characterized as it is by an increase in the number of things, by the extension of technology and "the increase of human beings on earth." Quantity as a factor of civilization, modern culture, cities: the dominant, prevailing culture within which—against which—the deculturized [dispossessed] must learn to survive. (1981, 134; Paul's brackets)

The oscillation of within-which-against-which has become a familiar pattern for us, a ritual. "Who has seen the wind?" sang

Yoko Ono, "Neither you nor I / But when the trees bow down their heads / the wind is passing by." We learn to survive our deculturization in overlay and passing-by as well as in what Cara Armstrong and Karen Nelson see as "carry over":

> Rituals are determined modes of action and interaction which can expand a person's relationship to the landscape and carry over time; past merges with an already obsolescing present and projects into the future. . . . As (re)w/riting and (re)reading, ritual can be used . . . to exploit the gaps within a system determined by the patriarchal hegemonic culture . . . [and is] a response to a genuine need on both a personal level for identity and on a communal level for revised history and a broader framework. (1993, par. 5)

The web is now the place of quantity in Olson's sense, and its quantity here too is increasingly termed an environment or society. As a ritual space we increasingly seek on the web some sense of Armstrong and Nelson's "revised history and a broader framework." Yet we are right to wonder whether in any sense (or in which of our senses) its ritual action offers the expansion of our relationship to landscape that Armstrong and Nelson argue carries (us) over time.

Though the verb for it is surfing, we rather wade into web, approaching as tentatively as someone's grandchild wades through bramble and approaches branch and vine and spiderweb. Though much is made (and marketed) of its search structures, the web is not yet a monument enough for us. We as yet lack intimacy with its places enough to know where to look. We are as yet only at the first stages of its overlay, and our searches are thus repetitions like waves. These waves too are marks of mind and fall into a ritual pattern of what we might call confirmation, disclosure, and contiguity. We approach the space of the web as water and reach into its shallows and its depths. Sometimes we reach into this space seeking merely the confirmation that one or another part of the world/body is here too, whether a list of species of birds or a tea merchant's inventory of mountain tea. Other times we seek disclosure, hoping to experience that an unanticipated part of the world/body is here, whether in the text of a poem about the wind or the homepage of a cousin. Once comfortable with this wavelike rhythm of confirmation and disclosure we seek the broader framework of contiguity, the changing pattern of smooth stones beneath an ever changing surface. In contiguity we confirm our sense that one or another part of the world is adjacent and contiguous from time to time by turns.

In this way the web transcends the inevitable spatiality of other hypertexts by becoming primarily ritual, nomadic, and ephemeral and thus also richly overlaid with our sense of space and time and body. By circling round our senses of confirmation, disclosure, and contiguity upon the web, we find ourselves moving from the shallows and dropping off into sense. A recognition of traversal prompts my Vassar student Samantha Chaitkin to offer "a brand-new metaphor" in her critique of Storyspace and other Cartesian hypertext representations:

> I'd rather . . . jump up into the air and let the ground rearrange itself so that I, falling onto the same spot, find myself somewhere different. Where am *I* going as I read? No, more where is the Text itself going, that I may find myself there. (1996, n.p.)

What is the place where we are if it is not the place where we think we are when we are there? Where is the text going if it is not the place where we are when we are on the network? Where do we wade and from what body? Where is the whir of the wind? (These short sentences form the pattern of an argument. Stones along a stream or seafloor.)

We are where we are. We fall into the same spot yet find its difference in ourselves. This cannot go without saying, and yet the apparent placement of the network, the puerile illusion of virtuality, tempts us to do so. The importance of our embodied placement, our actual reality, arises not despite but because of our increasingly networked consciousness. We are (again) called not to take leave of our senses but to repossess and renew them. "Only in a culture in which visuality dominates," says Heidi Tikka,

> is it possible to assign a reality status to a visual representation in which some sound effects may enhance a non-tactile, taste-

less and scentless world. Furthermore, the reality of the VR is not essentially visual, but . . . is made of a three-dimensional Cartesian grid in which the movement of the user can be traced as a series of exact coordinate points and which therefore locates the user as a punctual solid object among other solidly rendered objects. . . . The subject of the VR finds himself enclosed into a dataspace and deprived of a corporeal body. In the VR-space the abstracted vision becomes associated with a gesture. The pointing gesture moves the user forward in the constant state of erection. (1994, n.p.)

There is a sense of reading that mistrusts gesture, which is after all the work of surface, and thus demands to inscribe it in the mark rather than the gesture. Such mistrust is a maypole ceremony, its insisted mark a self-garlanding. Instead of this ceremony of erection I have in chapter 1 here characterized hypertext in terms of contour: "how the thing (the other) for a long time (under, let's say, an outstretched hand) feels the same and yet changes, the shift of surface to surface within one surface that enacts the perception of flesh or the replacement of electronic text." We are where we are and it is a mistake to claim, even in cyberspace, that we are anywhere else. "Interacting with electronic images rather than materially resistant text," N. Katherine Hayles writes,

I absorb through my fingers as well as my mind a model of signification in which no simple one-to-one correspondence exists between signifier and signified. I know kinesthetically as well as conceptually that the text can be manipulated in ways that would be impossible if it existed as a material object rather than a visual display. As I work with the text-as-image, I instantiate within my body the habitual patterns of movement that make pattern and randomness more real, more relevant, and more powerful than presence and absence. (1993, 71)

The memory of the world, materialized in the body, for which both the book and the screen stand as repeated instances of embodiment, is itself the world. This is the nature of the erotic, another name for mortality and our presence in a real world.

"Through ritual, individual, private actions can become part of a shared act," write Cara Armstrong and Karen Nelson.

> Through repetition, actions can take on additional significance. Repetition can enlarge and increase an idea or purpose and may also suggest eroticism. Rituals, as shared acts, are potentially inclusionary. Ritual layers daily experience with the cyclical and the symbolic. (1993, par. 7)

We are who we are. We are where we are. Layered and overlaid, we make a world within our bodies.

Water, or, The Body Again

The mediums of exchange for the skin are light, air, and water. We examine them in order to see ourselves as who we are.

Water is the figure of the body as a medium of exchange. There is the formless place where the world is made. Heidi Tikka lingers like water over the smooth stones of the "continuous, compressible, dilatable, viscous, conductible, diffusable" qualities that Irigaray describes in "Mechanics of Fluids." In that essay Irigaray notes how fluid "makes the distinction between the one and the other problematical: . . . already diffuse 'in itself,' [it] disconcerts any attempt at static identification" (1985, 111).

Mostly water ourselves, we are singularly plural and simply mindful of its complexity. In her own critical appreciation of Irigaray's notions of fluidity, N. Katherine Hayles observes how "within the analytic tradition that parses complex flow as combinations of separate factors, it is difficult to think complexity. . . . Practitioners forget that in reality there is always only the interactive environment as a whole" (1992, 21).

We read ourselves in ebb and flow within the whole of water. The space of our mortality is the singularity of water, which turns by turns from solid to liquid to air. We mean within a flow of meanings, ourselves the repeated eddy of erotic gesture, ourselves the screen that, in Alice Fulton's phrase, "oscillates under consideration," ourselves as well the moist and knowing eye, a flow over the skin or pulp of the page. We ourselves likewise mark and mean the repeated touch of surface to surface within one surface, cyclical and symbolic, which enacts the perception of flesh. Beyond next before us once again we ourselves discover the current flow of electronic text within a desert of silicon. In not yet published speculations, Alison Sainsbury 1994 considers the reader as the literal (I am tempted to write littoral) site of inspiration, breathing out breathing in, and thus casts the act of reading in terms of lung or gill, the membrane and surface of vital exchange. She insists she means no metaphor but a cognitive theory; meaning is an exchange of moisture. Our selves and our cells argue as much.

A similar exchange prompts Carolyn Guyer to conceive meaning in terms of an estuary that "at any moment

> contains some proportion of both salt water and fresh, mingled north then south then north again by the ebb and flood of . . . tides. Right here is where I am.
>
> . . . The present is a place as much as it is a moment, and all things cross here, at my body, at yours. It is where I consider the past, and worry about the future. Indeed, this present place is where I actually create the past and the future. (1996a, 157)

She insists she means no metaphor but an actual ontology: "Nature is what we are, and so cannot be opposed to, or separate from, humans and their technologies." What comes beyond next is likewise inseparable and nothing less than what is before us: ourselves as expressed within time and space.

"When we get older," Sherman Paul writes about Robert Creeley's poem "LATER,"

> we especially want the comfort of intimate space
>
> Where finally else
> in the world come to rest—
>
> By a brook, by a
> view with a farm
>
> like a dream—in
> a forest?

We move toward the feminine, toward repose. We wish to enter the *gymnaeceum,* the house (always maternal) of all houses, that of our childhood. (1981, 62)

Because it is named as such, because it is cast as both a wanting and a wish, this space seems different to me (it is different) from Debray's protected and protective space where one "can take one's lodgings . . . so to speak, even curl up comfortably." Sherman Paul's space is explicitly not a return to room (or womb) but an older space beyond next before you once again.

This orientation was something he wrote about explicitly, elegiacally, himself coming round again to the fortuitously bracketed senses of doubled life with which he began his scholarly journey. In "Making the Turn: Rereading Barry Lopez," Sherman accounts the body's exchange:

It is salutary to divide the day between the work of the mind and the work of the body—the *vita contemplativa* and the *vita activa,* the latter, as I practice it, *menial,* according to Hannah Arendt—and it is necessary. The work of the body, outdoor work, is *out:* To do such work is a primary way of being-in-the-world, of finding oneself in the cosmos, in touch with things, physically "at home." The work of the mind, indoor work, is *in,* doubly interior: To do such work is often a way of withdrawing from the world, of living with its images. I use the spatial distinctions *(in/out)* that accord with the dualisms of *mind/body, subject/object, self/world,* but these are dualisms I wish to overcome: when *out,* by a participatory activity of mind; when *in,* by a meditative activity that seeks in words to hew to experience. (1992, 68)

Inevitably this meditation on Lopez turns to water, "to the natural relationships of the little-traveled upriver country." Lopez, Sherman says, has "undertaken an archetypal journey, a quest of the kind that distinguishes our literature,

The springs of celebration: How often have we sought them in childhood and a world elsewhere; how seldom in the heart of darkness. Yet isn't the significant aspect in this instance the extent to which [Lopez] has made ecological study serve this end? the extent to which he has gone *out* in order to come *in?* With him, it may be said, the discipline of ecology heals the psyche and the healed psyche serves the unhealed world.

Here he finds the anima and dances with her . . . This is told, appropriately, in the clairvoyant manner of dream or fairy tale, and it is recognized as such, as a mysterious occurrence whose moral meaning is nevertheless clear. *He dances and tells stories:* with these sacred gestures he celebrates the springs. (1992, 84–85)

Growing old with technology, neither opposed to nor separate from nature, we watch water as if it were a seduction. Here we find the anima and dance. Form forming itself. Form drawing us to ourselves.

The charms of hypertext fiction, my dancing stories and technology, are those of any seduction, the intensity of likemindedness, a feeling that the story (and its teller) somehow match the rhythms of the stories you tell yourself. The vices are likewise those of seductions. What you think you see as your own mind is, as always, another's. Things pass. The links are like comets on the surface of a pond, doubly illusory.

What comes next? Will the web supplant or supplement the world or book? When we get older we move toward the feminine, toward repose. I'm a little tired of the supplant-and-supplement question (even if I am in some sense guilty of forwarding it). Linear and hypertextual narratives seem a polarity but are only opposite shores of a stream. Our literacy is littoral. There are no linear stories, only linear tellings or readings. *Supplant* is a strange word (the dictionary renders it in terms of "intrigue and underhanded tactics"); I prefer *succeed,* with all its senses. If the linear narrative, insofar as it is aware of itself as a form, has always wished to succeed itself (as it seems, at least by the witness of its practitioners, it has), then it is unlikely that the hypertextual narrative will be any less ambitious.

Water does as much as it travels or eddies, changing change, successively taking the same form. What comes next? What next literacy, what next community, what next perception, what next embodiment, what next hope, what dance, what home, what next light.

We will have to watch. "It is through the power of observation, the gifts of the eye and ear, of tongue and nose and finger," Barry Lopez says,

> that a place first rises up in our mind; afterward, it is memory that carries the place, that allows it to grow in depth and complexity. . . . we have held these two things dear, landscape and memory. . . . Each infuses us with a different kind of life. The one feeds us, figuratively and literally. The other protects us from lies and tyranny. (1990, 188)

We will have to watch. Consoled by a belief that nature is what we are, it remains to be seen whether we can move enough toward the "clear space" that Barbara Page locates within "the conscious feminism of the [experimental and hypertextual] writer." What "animates her determination" is "not simply to write but to intervene in the structure of discourse, to interrupt reiterations of what has been written, to redirect the streams of narrative and to . . . clear space for the construction of new textual forms more congenial to women's subjectivity" (1996, par. 26).

As artists and thinkers and teachers we long for animation, interruption, redirection, and construction. What comes next is before us, in landscape or memory alike. What scours the clear space are the waters of repossessing and renewing. Ever afloat in a journey to the place beyond next, we begin to settle for transient miracles before us once again, whether the truth of crows on an Easter morning or the lines that end Sherman Paul's *The Lost America of Love* and this essay as well:

> We must go back to sets of simple things,
> hill and stream, woods and the sea beyond,
> the time of day—dawn, noon, bright or clouded,
> five o'clock in November five o'clock of the year—
> changing definitions of the light.

Chapter 6

Songs of Thy Selves: Persistence, Momentariness, Recurrence, and the MOO

Until recently, when its host MOO, Brown's Hypertext Hotel, went down for renovations and redesign, you could find the following in the Hi-Pitched Voices wing of that space:

Anne's Work Room

You see a large sun-lit room looking out over a rambling English garden. The windows are open and the smell of honeysuckle wafts in on the warm spring air. There are three large wooden desks. Two are covered with half-finished bits of code, books, papers, jottings of stories and poems, pictures. One is kept clear and here, neatly stacked, is the current work-in-progress. This desk has a green leather top and several deep wooden drawers. It magically keeps track of anything written on it.

Obvious exits:
out => Hi Pitched Voices
You see scribbles here.
Anne is here.

The poet, hypertext writer, literary feminist, activist, teacher, and computer scientist Anne Johnstone, whose room this is, and who indeed from the objective claim of the language of this space still stands here midst her scribbles and the smell of honeysuckle, and whom I once hugged (tightly, in sweet regret for what might have been, bidding her goodnight after dropping her off at a Days Inn on a winter night in Poughkeepsie), died in her riverside cottage in Orono, Maine, in the company of her circle of women friends, strangers, hospice healers, and her Scots mother at a very young age, not yet forty-five, of a rapid and rapacious cancer on February 28, 1995.

I begin this writing in our garden overlooking the Hudson. There is in this sentence both a preliminary sense, as if something must happen, and an inevitable setting that most readers bring to it. The latter requires amplification: is it the Hudson in Manhattan, a place of high palisades, commerce, bustle? the Hudson of the Highlands, the green shoulders of Storm King and Breakneck opposite? or the placid Hudson below Albany, muddy and wide?

The amplification that the sentence requires is, of course, an aspect of the preliminary sense that something must happen. I am drinking a ginger beer in a brown bottle but also the sweet dregs of morning coffee in a robin's egg blue mug and tapwater from a fluted water glass.

I have amplified the setting inward, making a proximate geography. Not more than thirty feet from me a concrete Tibetan Buddha sits in a weedy sprawl of jade next to a bank of wild mustard along our neighbor's wire fence.

The connection between the preliminary sense and the inevitability is a mortal space. A much overlooked aspect of death is that it is, so far as we are given to apprehend it in our conscious lives, solely preliminary, in almost no sense (except the preliminarily elegiac) the inevitable closure we attribute to it. (A sailboat crosses through the gap between two houses, one of clapboard, another eighteenth-century brick, which is my vista onto this sunny June afternoon.)

We fill this mortal space with the living of our lives, although like a hole in the sand it is never filled or rather ever filled, since we do so in recurrence, wave upon wave.

It isn't necessary that the sentence be true in either the preliminary or the inevitable sense for the space of mortality to form. Fiction, of course, both accomplishes and fills this space, more or less finally, though the reader of a print text can see it as ever filling in repeated readings or memory, while the reader of an electronic fiction must inevitably see closure as transitory.

In like ways the space between two related ideas has its own mortality, its own preliminary and inevitable states, although these are not a polarity, rather a consistent relationship, a tensional momentum, in Carolyn Guyer's phrase (1996a, 161). Take two ideas thus far opened here: the idea of the tensional momentum of the preliminary and the inevitable and the idea of language as embodied proximate geography. One of them—ironically the latter—is less fleshed out. There is a claim here, hardly made as yet, that we live our deaths by somehow symbolically mapping space (through language, image, movement, even the implied spatiality of sound that creates through interval) in relation to our bodies. Death is a condition of suspended preliminarity. One table in Anne's Work Room is "kept clear and here, neatly stacked, is the current work-in-progress." This paragraph, as in waves it seems to close the gap between the two ideas that it isolates and emphasizes, might seem to be squandering the space it creates and describes (especially now as I take this prose to a more analytic and self-reflexive vein, although the appearance of a self in this parenthetical weave—an "I" who takes this paragraph "there" and who drinks ginger beer on a splendid Sunday afternoon along the Hudson—in some sense opens it again).

It isn't necessary that Anne Johnstone be present in her workroom for her character to continue to inhabit the mortal space between the preliminary and the inevitable. (Though I am truly in our unkempt and lush garden and there is birdsong—squawks, cheeps, bottle songs, and recurring trills—a jet overhead on its approach to Newburgh, powerboats along the river, the occasional melancholy and tentative song of baritone chimes when a fresh breeze climbs up the sunny incline to the sunbaked house where they hang outside our mudroom window, and as well occasionally the silence of sailboats tacking on edge or white moths fluttering in the mustard flowers when I look up.) In truth, as a MOO devotee, perhaps a techie impatient with my chimes and moths and blue mug, surely knows, someone (a wizard in early instantiations of MUD and MOO cores, though lately likely to be called anything: janitor, proprietor, citizen, concrete Buddha) could inhabit the Anne Johnstone character, revive her.

But it was ever so. These things happen. A letter arrives from a dead woman a day or week beyond her death. We revive her, see her in the mixed and mortal time between the preliminary and the inevitable, create for her a proximate geography (she held her pen so, sat at her computer in the wood-paneled cottage down a slope beside the river—I imagine this, I was never there, she could have written from the Days Inn, could be there still, children believe as much, the mad, the MOO devotee). I am not disinclined to agree with our devotee, the breeze springs up and magically the full sails of a Hudson River sloop, the *Clearwater,* appears rounding the point at the Bottini Oil Company tanks in New Hamburg. The baritone chimes are the Gregorian model from Woodstock Chimes. I recall a large sunlit room looking out over a rambling English garden. My name is Anne.

"In the future she wouldn't have to die, not really. When MOOs become graphical, where there's really VR, there could be an avatar of her. You could revive her, move her through the space," says the MOO devotee. "You wouldn't need language, words get in the way."

O this I will not agree with. We have to die, this is the meaning of words. Unpleasantly or no, I believe that in the interregnum of the MOO it is language that creates the proximate, mortal space between the preliminary and inevitable. It is probably my folly, certainly my garden (though rented, we own this space in observation and action, our landlords, the Cullens of Rabbit Island on the Hudson near here, are mere wizards). I have cast the MOO Devotee out, toaded him in MUD parlance (though why I have made the devotee a he, or why invited him here unless I needed his residence I do not know, unless, of course he is me, made in my form, cast from my garden: these are old myths). The concrete Buddha looks on serenely, his back to the river.

Words are the way, I want to say. (I rhyme like baritone chimes.) And yet I know, in the way someone watches water slip through sand, that words are being displaced by image in those places where we spend our time online; know as well that images, especially moving ones, have long had their own syntax of the preliminary and the inevitable. Time-based forms, whether music or filmic, depend on their inability to fill out the space they create and open.

Perhaps the MOO is more a film than a sestina (Anne Johnstone's form, she wrote a sequence of hypertext sestinas, what would seem an impossible form, and yet in this linked space every line seemed to double its sense of inevitability and recurrence, the frame filling with the possibility of the line, draining with the traversal of each link, all of it gathering again at the end in the eddy of envoy). But if a film, of what kind? Something like Godard's in the golden years, where the cast would arrive in the space of an idea, improvise, neither he nor they knowing where the camera went or what would follow this, what precede it. In an earlier chapter here I have written of the MOO that "attention is always elsewhere, it is a distraction, a disposition of self, the confusion of the space for the occasion." The actors join the audience in seeing the film as it results; the film as it results joins the director with the actors in seeing the film. This is the form of the sestina. (The MOO devotee, a techie, loses patience with this artsy talk and drifts off; he looks for action. He is soon elsewhere, in MOOspace where action inevitably follows the preliminarity of space.)

> You see scribbles here.
> Anne is here.

Against the commonly cited momentariness of MOO experience and the evanescence of the selves that form within it, there stands the rhythm of recurrence on unknown screens elsewhere; the persistence of certain "objects" that, like the consumerist flotsam of temporal existence (a brown bottle or a sailboat), mark the swell and surge of lives lived in body, space, and time; and the mark of the momentary itself, meaning within meaningfulness not against meaninglessness. Thus, like any poetic text, the MOO aspires to moral discourse and to inscribe our mortality.

Movement is the mark of the momentary, and our mortality, alike. We move through moments and thus mark them with our bodies. Alison Sainsbury appropriates Hélène Cixous's *Three Steps on the Ladder of Writing* to call for an electronic writing that moves beyond the "hypertextual body and its ability to disrupt and interrupt" to "the necessity of embodied text . . . moving under, around, not arriving . . . the perfect unity of textual body and jouissance" (1994, n.p.), which is to say one that inhabits the momentary and proximate space between the preliminary and the inevitable. The image Cixous has in mind in her text is the daughter of the Grimm fairy tale who each night pushes her bed aside, lets herself down through a trapdoor, goes off into the night and wears her shoes out dancing. "In order to go to the School of Dreams," Cixous writes,

> something must be displaced, starting with the bed. One has to get going. This is what writing is, starting off. It has to do with activity and passivity. This does not mean one will get there. Writing is not arriving; most of the time it's not arriving. One must go on foot, with the body. One has to go away, leave the self. How far must one not arrive in order to write, how far must one wander and wear out and have pleasure? One must walk as far as the night. One's own night. Walking through the self toward the dark. (1993, 67)

The character on the MOO is always starting off, must always "begin to begin again" in William Carlos Williams's phrase. Late in a MOO session someone joins and, despite whatever coherence the evening and the space has offered us, we are all once again starting off, not arriving where we began to think we were. Not just the character (our own or this late-arriving one, perhaps it is the MOO devotee coming back; perhaps it is the avatar of the concrete Buddha as Anne Johnstone) offers us a lasting presence of particularities as a strategy against the fragmentary plenitude and multiplicity that faces and effaces us. We are all just starting off, all recently arrived. Someone moves from another room to this one—one of those characters who arrives in a comet of textual introduction, dragging a tinker's bag of objects and attributes behind him as the tail (or the tale: it *is* the MOO devotee!)—and his arrival disrupts the surface of what we have settled on among us (as the comet light stirs a dark pond). His presence once again marks the momentary space between the preliminary and the inevitable. He is what catastrophe theorists call a perturbation. (Oh here let us free him from his gender and his character, we are in a MOO after all; he is not the MOO devotee, she is the shapeshifter.)

What can happen now that she is here (not yet arrived at the dance and yet here)? We can summon hierarchical notions of meaning and power: toad her, burn her at the stake, ignore her (Judy Malloy in the early stages of Brown Kitchen would set up space inside a room at Lambda MOO and begin to tell her stories, ignoring the yelps and noise and already flowing meanings until the story made its own space, like the eddy of a contrary flowing river). Or we can (continue to) make our own meaning within and against and in complicity with the arriving comet. Yet in the shifting current the eddy (the mortal space between the preliminary and the inevitable) has both enfolded and unfolded. What we make will never be the space of what went before; what went before was never the space we thought it was as it unwound, scrolling upward like smoke, flowing away like a river or towns on a Triptik.

The arrival of the avatar (shapeshifter, Judy Malloy, Anne Johnstone, our friend the MOO devotee) offers an occasion of coherence. Coherence in this sense is very close to what the catastrophe theorists (characters who arrived not long ago in a comet's spew) mean by a singularity or phase shift, that is, a recognizable change in which something amorphous takes on form defined by its own resistance to becoming anything other than its own new form. Coherence is making sense for oneself and yet among others.

A room in a MOO is not the meaning of what it contains but the space of its coherences. You choose to understand the flow of talk and action that you are within, choosing also from what perspective (what self) to view it. Following Cixous, Alison Sainsbury suggests that we must go beyond the self in order to reclaim the active self as something newly whole. Coherence can be seen as partially meaningful patterns emerging across a surface of multiply potentiated meanings into something newly whole: the eddy in the river, a brown bottle, a desk with a green leather top that "magically keeps track of anything written on it." Coherence distinguishes itself not against but within other possible coherences, in the recurrent flickering of meaningfulness not meaninglessness.

You see scribbles here.
Anne is here.

Days have passed, it is (Tuesday) too steamy to sit by the river, although the Buddha has not moved. I want to walk out after dark, a walk through the self toward the dark.

Let us talk of movement for awhile and see where that gets us. I have often made the claim that the MOO, as a variety of hypertext, is essentially spatial. Yet how to reconcile this with the contrary and obvious claim that the MOO is temporal? What we are looking for here is the mark of the momentary, the space between the preliminary and inevitable. Recurrence, I have said elsewhere, is the memory of spatial form. How something can mean, not what it does mean. The repetition of a phrase, "Anne is here" for instance, marks both an indeterminate space (unless you happen idly or paranoically to have decided to count the lines or paragraphs or pages after a certain phrase in case it should appear again) and an indeterminate time (you might likewise have either used a stopwatch, or looked for her, or decided this phrase is a narrative: first she was there, then she was reported dead, now he says she is here but means something else by it; this succession of temporal measures—metric, experienced, narrative—is the stuff of the MOO experience). The MOO repeats itself both determinately and indeterminately. The rooms are (largely) determinate, characters less so. The formulaic aspect of many actions (so-and-so says such-and-such) are determinate, the substantive (Devotee says, "This is philosophical claptrap") are not.

Yet these distinctions do not sustain themselves. Moving from room to room in some MOOs (Hi-Pitched Voices for instance) we accrue meaning both directly (reading the walls as it were) and indirectly (from the customary interchanges among characters). Rooms thus oscillate between determinate and indeterminate states according to the order in which we encounter them and in whose presence (or absence: Anne is here) we do so. Likewise actions, which seem most indeterminate (Devotee snarls, scorns, turns into Shapeshifter; Catastrophe Theorist rains and rants and reigns) become by their recurrence markers of a constant set of changes that cohere as the character who enacts them.

Even the measures of temporality oscillate, as in Einstein's famous trope of the relativity of time for someone seated on a hot stove versus someone seated on a lover's lap where experienced and metric time interact; or again how in their metrics some video games observe narrative time (or what we could call Disneyland time, at three-quarters scale or less), compressing "thirty minute" halves of a soccer game into twelve. The coherences of spaces, characters, and temporality while simultaneous are not always contemporaneous or coterminous in the MOO. Not just the MOO experience but its coherences (what we might call "the meaning") are distributed, duplicated, alternately experienced across the screens of its participants as if an audience watched a film projected kaleidoscopically.

Perhaps the MOO is more a film than a sestina, I suggested. Interactive filmmaker Grahame Weinbren (1995, n.p.) considers the question of a similar system of indeterminate narratives and explorations. In the "interactive cinema," he says, "All and any loose narrative ends will never be knotted. . . . If a viewer navigates through a mass of material, some of it will be seen and some won't, and surely some of what isn't seen earlier will raise issues that remain unresolved in what is seen later."

How what is not seen might raise issues that we later recognize as unresolved rehearses the connection already made here between the preliminary and the inevitable within a mortal space. We all die of what at some point were unknown causes and in so doing resolve what we did not see before. We attribute meanings to the cessation of experience, this is the nature of closure. As a slogan we might say that moral resolutions emerge from mortal recurrences. That is, we attribute morality to the mortal space between the preliminary and the inevitable.

There is a mechanism for this, although we are unlikely to attribute to it the cultural certainty with which certain narrative theorists attribute a moral frame to the beginning-middle-end coherence. "A system can be sensitized to repetition," says Weinbren, "either so as to avoid it, or so that as soon as repetition starts the viewer is offered the opportunity to enter a structurally different region, a territory of culmination or summary." People are not systems (systems are what people are not) and yet we do program ourselves to avoid repetitions (it is Wednesday now, back in the garden, and across the road there is the sound of a gas-powered trimmer that I ignore, nearer a blue jay's hoarse, repeated, jug-deep song); do consider that some recurrent experiences (the seven danger signals of cancer for instance, or love) signal a territory of culmination, that is, ever after perhaps happily.

However, while it is true as Weinbren says that "in general terms, a map of territory covered can be kept by the system, and once a certain area has been explored, closure possibilities can be introduced," this seems where, despite our most fervent, earnest protest (and, Horatio, despite all your cognitive sciences), people and systems differ. We don't "get it." Don't "work our way through" the territories of the heart until we reach a place where suddenly we can "see clearly," a clearing where we know "what's what, what's up, what matters."

We do, of course (do I contradict myself? well then I contradict myself), but at the level of longing, of desire. We fill the mortal space between the preliminary and the inevitable with a moral landscape of our own (re)making, a succession of coherences (as it was in the body, is in the MOO, and ever shall be in memory). The moral landscape remains a Hesiodic space, the world (even the MOO, the blue jay, or the concrete Buddha) is held by the glue of desire. "All this is to say," says Weinbren, emerging to a clearing where he can see what's what, what's up, what matters, "that despite its need for an opened narrative, closure cannot be banished from the Interactive Cinema. Remove the imminence of closure and we begin to drain cinema of desire. Closure must be recast in a more radical light."

So we shall. Let us consider the most radical light, the light of lights, and look on the moral landscape with a theologian's eyes. There (or at least in this one theologian, Gabriel Marcel), the truth is that the dead live, Anne is here, you see scribbles:

> The spirit of truth bears another name which is even more revealing; it is also the spirit of fidelity, and I am more and more convinced that what this spirit demands of us is an explicit refusal, a definite negation of death. The death here in question is neither death in general, which is only a fiction, nor my own death insofar as it is mine . . . ; it is the death of those we love. They alone are within our spiritual sight, it is they only who it is given us to apprehend and to long for as beings, even if our religion, in the widest sense of the word, not only allows us but

even encourages and enjoins us to extrapolate and proclaim that light is everywhere, that love is everywhere, that Being is everywhere. (1963, 147)

It is a commonplace to think that MOO pranking, MOO terrorism, MOO marauding, and worse (see Dibbell 1993) hearken back to the days (they continue of course) of the MUD and its roots in adventure games and especially in D&D, where boys killed boys in the guise of knights and dragons and orcs and theologians. That this commonplace happens to be true does not keep us from recognizing that the MOO is a place of death's denial, both in the commonplace slaughter of imaginary (though no less dangerous) boys and the riverside workroom of the poet as well. Marcel's notion of denial as fidelity, while couched in negation, is imbued with a positive charge of desire. According to Marcel our task (our life) is "to apprehend and to long for" the dead "as beings."

It isn't immediately obvious, even on the MOO, how to engage in a language of denial and desire. You can't simply say the dead live (Anne is here), nor is Marcel's theology an abracadabra one in which we do so. Fidelity requires repetition and recurrence, it is in this sense only "true to life." True *to* life is true *through* life (this is another, equally sloganeering, version of the earlier slogan "moral resolutions emerge from mortal recurrence"). As Cixous suggests (I have long thought that at some early age she read Marcel, based on little more than intuition and certain echoes in their language), "We live bizarrely clinging to the level of our age, often with a vast repression of what has preceded us: we almost always take ourselves for the person we are at the moment we are at in our lives" (1993, 69). This is a misappropriation, a reversal of the relation between the preliminary and the inevitable, seeing oneself as arrived rather than not arriving. Cixous, longing as always, seeks to reverse the reversal: "What we don't know how to do is to think—it's exactly the same as for death—about what is in store for us. We don't know how to think about age; we are afraid of it and we repress it." What is in store for us is another way to describe the preliminary, we think about age by marking the momentary, through recurrence. "Writing," says Cixous, "has as its horizon this possibility, prompting us to explore all ages" (1993, 68).

Here a self-satisfied MOO devotee could stop, note (even exult) that MOOspace is little else than a horizon of possibilities, an exploration of seven times seventy ages of woman and man from Jesus to Jacques to Janis (or Scott) Joplin, from Muhammad to Melanchtha to Mrs. Ramsey. Yet we likewise need to stop here and consider what seems the evident drabness and regular debasement (as in the smart-ass sequence from Jesus to Joplin) of MOO language. The debasement of language in electronic texts leads some cultural commentators and critics to argue that theory such as this essay overstates the ordinariness or worse of the language of MOOspace, the web, and so on. In fact, canonical critics might argue that the argument here is carrion criticism, parasitically feeding on waste in order to puff up a claim of transcendence and the poetic. The MOO (all of electronic text) is faulted for not having a language worthy of eternity. This is of course a macho claim, electronic space doesn't have balls not brass but *aere perennius* (more lasting than bronze in Horace's phrase) enough to stand up to history. Canonical criticism is a claim made by imaginary boys not different except in their pedigree than MOO marauders.

Such challenges are not so much overthrown (it is the same game, a joust of imaginary knights: these canonical critics are straw men, in my own image and of my own making as much as the MOO devotee) as much as seen through, both in the sense of persistence and the transparent sense of boys in costumes. It isn't necessary to see the MOO, as Erik Davis does, as the "apotheosis of writing . . . nonlinear texts in many ways more marvelous than the precious literary experiments" (1994, 43). Instead we can see MOO writing as something much more routine and commonplace (what is more commonplace than imaginary nights? more commonplace than desire and denial, than religion?). Cixous in the "School of Dreams" section of *Three Steps on a Ladder of Writing* sees the commonplace "face of God, *Which is none other than my own face,* but seen naked, the face of my soul" (1993, 63). Cixous's face of God is a stack of tender lies, what in speaking of electronic text my student DeAna Hare calls "kitchen talk." The recurrent mark of the momentary (ungendered, though most common among women) becomes, however briefly, a coherence for Cixous:

> The face of "God" is the unveiling, the staggering vision of the construction we are, the tiny and great lies, the small nontruths we must have incessantly woven to be able to prepare our brothers' dinner and cook for our children. An unveiling that only happens by surprise, by accident, and with a brutality that shatters: under the blow of the truth, the eggshell we are breaks. Right in the middle of life's path: the apocalypse; we lose a life. (63)

It is exactly in the commonality of MOOspace, the noise and actions, the kaleidoscopic projections, the constant replacement that I have suggested is the characteristic of electronic text, which, rather than in the text itself (the content, to use that reprehensible word with which technocommerce would end art) that we lose and gain our lives. MOO talk is kitchen talk, a gathering of aunts in which, as Grahame Weinbren characterizes the intermingling narratives of Interactive Cinema, "several story lines continue until one, some, or all of them end" (1995, n.p.). Weinbren too begins with water (Rushdie's *Haroun and the Sea of Stories* and Barth's *Tidewater Tales*), bringing him to a place where "numerous Diegetic Times are constantly flowing forward, many narratives operating in time simultaneously whether or not the viewer encounters any particular one . . . potential narrative streams, elements themselves unformed or chaotic, but taking form as they intersect, gaining meaning in relation to one another" (1995, n.p.).

Let us end then in water as well and (for awhile, until the end's end) a series of dead white men, flickering by as if on a screen of light. I suppose one could begin to end with Eliot (I have regularly been accused of being a modernist) whose river in the *Four Quartets* is "a strong brown god . . . ever, however, implacable / Keeping his seasons and rages, destroyer, reminder / Of what men choose to forget," and who "tosses up our losses" (1943, 35–36). But instead, longing as always, let us heed the call of desire and denial (do you wonder where this hortatory "us" has entered? or did you, determinate, mark the paragraphs since I turned to this diction, creating you in my own image? It is a proximate geography: under a lapsing morning haze the river hesitates ripe and fragrant between tides, not flowing either way, as a breeze strikes up out of the humid overcast and the rhythm of the birdsongs hastens, the jay turned now from jug-song back to the more ordinary grating call).

Against the commonly cited momentariness of MOO experience and the evanescence of the selves that form within it, there stands the rhythm of recurrence on unknown screens elsewhere; the persistence of certain "objects" that, like the consumerist flotsam of temporal existence (a brown bottle or a sailboat), mark the swell and surge of lives lived in body, space, and time. The mark is the mark of the momentary itself, meaning within meaningfulness not against meaninglessness. The flicker on so many screens elsewhere is denial and desire together, the oscillation of the preliminary and the inevitable, a serial recognition (temporal, recurrent, harmonic because at intervals) of our shared mortality and persistence. In "Crossing Brooklyn Ferry," Walt Whitman (1965, 159) wrote:

> It avails not, time nor place—distance avails not,
> I am with you, you men and women of a generation, or ever
> so many generations hence
> Just as you feel when you look on the river and the sky, so I
> felt,
> Just as any of you is one of a living crowd, I was one of a
> crowd,
> Just as you are refresh'd by the gladness of the river and the
> bright flow, I was refresh'd
> Just as you stand and lean on the rail, yet hurry with the swift
> current, I stood yet was hurried.

Writing, Whitman puts us at Cixous's horizon of possibility, "prompting us to explore all ages." This is a conscious attempt at a proximate geography, a claim for the transcendence of the virtuality of language over the mortality of the body. The claim for transcendence lies as much in the recurrent actions of the "you" (it is us) who are addressed and summoned to the spirit of fidelity as it does in the language itself. We persist in the mark of the momentary, the endless starting off of denial and desire alike, the commonplace and anonymous rhythm that Samuel Beckett summons in "From an Abandoned Work":

> Oh I know I too shall cease and be as when I was not yet, only all over instead of in store, that makes me happy, often now my murmur falters and dies and I weep for happiness as I go along and for love of this old earth that has carried me so long and whose uncomplainingness will soon be mine. Just under the surface I shall be, all together at first, then separate and drift, through all the earth and perhaps in the end through a cliff into the sea, something of me. (1995, 160)

Anne is here. Let us end then. I write this in preliminary apprehending and inevitable longing for Anne Johnstone who lives, no less in these words than in her embrace impressed once upon the body of my memory.

One Story: Present Tense Spaces of the Heart

There is one story and one story only.
  —Robert Graves

The struggle for language and the struggle against perfect communication, against the one code that translates all meaning perfectly . . .
  —Donna Haraway

As I get older I am convinced I have only one story and that it is multiple. The forms of things mean.

  as much as.
  in the same way.
  (Do not fill in the blank)
  (There is no blank.)

There used to be a colon in the first sentence. The colon in the first sentence formed a polarity, or so I thought. (Donna Haraway: "the relationships for forming wholes from parts, including . . . polarity and hierarchy are at issue" [1991, 177].)
  There is no blank.
  When I was younger I was convinced of a story that became for me multiple. When I was younger I was a story that became me multiple. When I was younger I read the making of americans, the maximus poems, a poet in new york, the blood oranges, each one beginning with an article. (Do not fill in the blank.) Indefinite, definite, infinite.
  It is not the comparative, the whole (from) part, that is the missing limb from the first sentence, it is rather the report of the action of the thing, the how. I/know/nothing/new. Olson in "GRAMMAR—a 'book'" (1974, 27) charts it, typographically—indefinite, definite, infinite, who, how, like, as, a, all:

BODY . . .
of *lic* (who-like?)
of what sort or
kind

Among the Mexica the good scribe, it is said in the Codex Florentino, "knew very well the genealogies of the lords" and of him it was said, "He links the people well; he places them in order." Sister stories know a different order, they link not by placing but by finding places within which to be. They know very well what, in *The Making of Americans* Gertrude Stein called "all the kinds of ways there can be seen to be kinds of men and women" (1934, 73). Hypertext, a linking technology, tells of (and with) different orders. Carolyn Guyer, Rosemary Joyce, and I have collaborated on a multimedia hyperfiction *Sister Stories* (1999) that explores ways to be women and men. Building from the mythological story of Coyolxauhqui, sister to Huitzilopochtli, the text itself explores the nature of telling and of reading, of being inside and outside a story, a place, a field, a history, a text. That is, how things mean or/as the body: one story: multiple.

For the Nahuatl-speaking Mexica (whom we have come to know as Aztec) the boundary of text and image was artificial. They wrote in highly pictographic ways, providing specifics of when, where, who in text, and specifics of action through images. Similarly in hypermedia as I have said elsewhere the image again takes its place within the system of text, the word again takes its place within the universe of the visible and the sensual. For the specifics of action in Storyspace (our hypertext software) the unit of linking is at the moment a four-pixel square (the smallest possible "region" link in a graphic) or the single letter (character) in text. The formal organizations, how the forms mean, are multilevel networks across an underlying hierarchy, networks that (despite their primitive box-and-lines graphics) some writers exploit as visual-semantic entities and some readers claim as the emergent, dynamic form of the texts they create in their readings. In transforming this work for the web we will have to approximate in surface and shifting surface this more vivid and dynamic network. In *Sister Stories* the word as image and action is meant to implicate readers and writers alike. The collaborative reading unfolds on the assumption that "in a history of many men and women," as Stein says, "sometimes there will be a history of every one" (1934, 396).

The forms of things mean. The story my archaeologist sister sees in this story of sisters is formed of shards distributed across the fourteen volumes of the codices. Likewise the shards of worn glass washed up on the beach in the novel by my heart's sister, Carolyn Guyer:

> Seeing the afternoon's harvest there in a pool on the bed made her catch her breath. These shards looked so much like the Lake to her, the angles and curves of the surface beat, sharp and dull, blue green umber light merging, blending, overlapping. As with the Lake, she could not take her eyes away. She had with her a square, shallow cigar box, bought on a whim for its bright, ornate decoration. She'd been keeping pencils and odds and ends in it. Removing them, she poured the beach glass into the box. The inside of the lid was golden and hot as midday. She opened it wide, and the color of the Lake shifted musically beneath it. (1992b, n.p.)

What's found is not buried there as code but rather shines through: not Hansel's strewn breadcrumbs but rather Gretel's persistence of vision against the coming night. Light as transmission, the forms of things. Decades of Mexica informants responding to the old Franciscan Sahagún's proto-ethnographies successively learned from these questions themselves how to tell him the stories in the forms his culture could hear. My sister points out in the codex illustrations how the figure of the madwoman—she is merely unmarried and so a threat, a storyteller—is moved out into a landscape the Spanish will recognize, the pictographic and narrative space constructed by negotiation: the woman pictured sitting in an open landscape, isolated and iconic as an exotic bird. Likewise Dwight Conquergood explained to me how in the refugee camps the Hmong constructed a new aesthetic for their *paj ntaub* ("flower cloth"), whether adapting encoded motifs of stitchery to more pictographic (comic strip panel) camp narratives complete with embroidered captions, or on the other hand churning out Peaceable Kingdom tree of life scenes, in each case shifting from traditional psychedelic circus colors to muted yuppie tones of blue and gray—all to suit a market they learned about in camp stories of blooming dales and Pier One. Yet the stories remaining there still for one who traverses the encoded stitches, she who unearths the different shades within.

Rosemary Joyce (in email, what we used to call—the templates, MLA, or Sahagún—"a personal communication"):

> Blue and green are without a doubt the most prized and significant color(s). Technically, the Aztecs did not separate as we do the entire blue-green range. So all those color terms are associated—with water, vegetation, growth. (Green feathers are described as like new plants sprouting, and in some Mesoamerican cultures stand for the springing forth of new children.) Of course, like all humans, Aztecs could see the difference between different shades of blue-green. They made these differences concrete by using gem names for them: turquoise, xiuhuitl; jade, chalchiutl. Blue then was associated with sky, and with sacred fire. Jade with vegetation and sweet waters of the earth.

There is a language of feathers, a language of glass, a language of stitches, a language of contours, a language of story spaces, a language of each of us and of every one of us. You cannot describe it, you can only see it (in memoriam is not postmortem, however avant). "That's why," Charles Bernstein says (aptly) in *Content's Dream,* "That's why I object when you say it makes sense 'to me' as opposed 'to you.' Why talk about making sense *to* anybody and not *of* something" (1986, 420).

We are trying to see a truly participative, a multiple, fiction and so for the moment at least peer into the ultramarine depth of the computer, our eyes moving over the flickering text as snorkelers along the coral reef, floating there without the anchor of the book. As in reading (with/in) the body, Olson's proprioceptive, or as Janet Kauffman has it,

> Underwater, things are not what you think. There is no confusion, first of all, no one-way traffic, no solid or dotted lines, your sense of direction is exact, irrefutable, whether you go with the

current or against it, whether you cut at right angles to it, or sit stock-still, footed in mud, clammed shut. This is the lost and found knowledge, the assurance of touch, head to foot. This is buoyancy, hazard, and waywardness—what it is to be at home, unhoused, ongoing. Elsewhere, alert, you have to admit, nothing surrounds you, not air, no, it's out of hand. We've pushed it off, walled it in, walled it out. However it once was for a person's body, moving around, when it was creaturely, thoughtless, is a recollection that comes back only in lapses, when we lose track—in lovemaking when it is ranging, sweated, benumbed; or when we fall on the ground as children, dizzy from spinning, and feel the ground under us careening. (1993, 61)

Indeed I conceived the "words that have texture" ("words that yield") in *afternoon* as something of the "recollection that comes back only in lapses, when we lose track," discovered in places where the reader literally could press against, caress, tease out (extrude) the language. Yet the aim of hypertext (multiple) fiction is one story (the story of its own telling), "the assurance of touch, head to foot." How, as I noted in an earlier chapter, Erin Mouré says, "the reader has to switch too, has to be prepared to say, 'Okay there's more than one kind of reading'" (1993, 38).

Okay there's more than one kind of reading. A similar realization propelled my oft-quoted impetus for *afternoon* toward "writing a story that would change each time you read it." The forms of things mean. My interest has always been on the development of multiple narratives as whole forms (i.e., neither self-reflexive nor ironic, not exhausted possibilities), but where the whole exists in something like Donna Haraway's situated knowledge (*"simultaneously* an account of radical historical contingency for all knowledge claims and knowing subjects . . . *and* a non-nonsense commitment to faithful accounts of a 'real' world" [1991, 187]) as present tense spaces of the heart.

Its insistence on the wholly formed multiple reading makes hyperfiction's connection with so-called postmodernist fiction seem to me unfortunate. Postmodernist (fabulist, metafictional, etc.) in that sense suggests flatness, lackluster bubbles from day-old champagne. (Avant at least restores the pop [shubopshubop] over black ice.) *Postmodernist* is as inadequate to describe hyperfiction as it is to describe Robert Coover's claims for how form means: "I'm much more interested in the way that fiction, for all its weaknesses, reflects something else—gesture connections, paradox, story" (1983, 69).

Yet now that the mirror's gone what's that reflection of something else? What seems to me most interesting about hypermedia now is the inherent tensional opposition (to use Carolyn Guyer's phrase) between image and text. Haraway similarly identifies "a map of tensions and resonances between the fixed ends of a charged dichotomy . . . [in which] knowledge [can be] tuned to resonance not dichotomy" (1991, 194). This calls to mind what Coover has called the "vibrant space between the poles of a paradox . . . where all the exciting art happens" (1983, 69).

What you cannot describe but you can only see is the inverse of the simulacrum, the real for which there is no fixed representation, or what in the first chapter here I've called the contour:

> how the thing (the other) for a long time (under, let's say, an outstretched hand) feels the same and yet changes, the shift of surface to surface within one surface that enacts the perception of flesh or the replacement of electronic text.

It's this surface, how the forms of things mean, that makes me less willing to throw over (in memoriam is not postmortem, however avant) the other postmodern, the one we've been summoned periodically to bury not to praise. Or at least not the postmodern of what Jane Yellowlees Douglas has called "the genuine postmodern text, rejecting the objective paradigm of reality as the great 'either/or' and embracing, instead, the 'and/and/and'" (1991, 125).

Avant-pop, and *après nous,* is always (and, and, and always will be) situated where Douglas situates hyperfiction, that is, a space in which the "third or fourth encounter with the same place, the immediate encounter remains the same as the first, [but] what changes is [our] understanding" (1991, 118). And so avant-pop seems to me the shift from carbon copy to a process of carbonation in both its (dictionary) senses (American Heritage indeed). In the first sense a wellspring (Canada Dry: a virtual landscape) reenergized with sparkling industrial bubbles, colorless as diamonds, soft as pixels; in its second, darker sense nonetheless the carbonized mark left behind by life, pencil or Hiroshima's (for example) ineffable, horrific, cataclysmic (and/and/and), late mil-lennial imprint, fossil's scream as virtual lifescape. We are sur-rounded by the marks of ghosts, the face that shines through the charred coals, the flesh of Aliquippa hills gouged by the 737, in each case writing in light, the technology not making us any less, in Rothenberg's fine title for it, *Technicians of the Sacred,* that is, "Within this undifferentiated & unified frame with its open images & mixed media, there are rarely 'poems' as we know them—but we come in with our analytical minds and shatter the unity" (1968, xxii).

I flew out of Pittsburgh by chance on September 8 at exactly the same time, more or less (7:10 P.M.) that USAir 427 went down. I saw nothing. I saw the whole thing.

> [Among] the ways in which primitive poetry & thought are close to an impulse toward unity in our own time, of that the poets are forerunners [is] . . .
>
> 4) an "intermedia" situation, as further denial of the cate-gories: the poet's techniques aren't limited to verbal maneuvers but operate also through song, non-verbal sound, visual signs, & the varied activities of the ritual event: here the "poem" = the work of the "poet" in whatever medium or (where we're able to grasp it) the totality of the work. (Rothenberg 1968, xxiii)

Outside the hills around the city ran with a gold river of houselights twinkling like a bedded fire below the gray-blue ridge of pumpkin-edged clouds as the sun set. Just ahead the horizon glowed melon, copper, cantaloupe fading to a paler blue-gray sky above, etched upon the cocoa hills and horizon's edge below. Off through the gap of gilt gray clouds Venus lay in the moon's thin, white arms (as the world-mapper Van Allen long ago said it, calling us all out from the party into the Iowa City night to see).

> "Why don't you write," I say, "a sentence like that?"
>
> Margaretta refuses to write. "I read," she says, "I'll read the sentence."
>
> Her desire, she says, is to leave no trace. (Kauffman 1993, 81)

"There was a hand with a ring on it," a minister said in the wire service story, "they think it was a stewardess because her uniform was nearby." It will not do to correct his terminology. No longer a flight attendant, the trees strewn with awful fruit beneath a cantaloupe dusk. "When we arrived," a rescue worker said in the same story, "there was no one to save."

What we read in the difference between the desire and the trace is how the forms of things mean, more than one kind of reading, the totality of the work, the inverse of the simulacrum, the contour, the multiple story.

Chapter 7

Nonce upon Some Times:
Rereading Hypertext Fiction

That which is reread is that which is not read. The writer rereads and unreads in the same scan, sometimes looking for the place that needs attention, other times seeking surprising instances of unnoticed eloquence that her attention now confirms in a process of authorship. Most often she looks for the thicket, the paragraph or phrase that relinks a vision or reforms it, a vision that she put aside or lost, that dwindled or lapsed, that exhausted her or she exhausted. In the process of reading for what she has not written (or written well) she often does not read what she has written well (or not written).

For over eight years now in workshops with writers exploring hypertext fiction I have posed a question about rereading and held my breath fearing an obvious question in return.

Suppose at this point your reader, before going on, has to reread one part of what comes before, I ask.

No one asks why. There are reasons.

For the writer rereading, the question seems to be one of ends. "What happens at the end of a text?" asks Hèléne Cixous; "the author is in the book as we are in the dream's boat. We always have the belief and the illusion that we are the ones writing, that we are the ones dreaming. Clearly this isn't true" (1993, 98). While Cixous is not thinking explicitly of hypertext here but rather the novels of Thomas Bernhardt, she nonetheless evokes the reader's experience of hypertext. Hypertext only more consciously than other texts implicates the reader in writing at least its sequences by her choices. Hypertext more clearly than other texts seems to escape us before we have it formed into an understanding we might call a reading. It beckons us as it escapes. The writer reading (or the reader as writer) thinks toward ending but more often looks for transport and escape, a way out that is after all another way in. It is as Cixous says:

We are not having the dream, the dream has us, carries us, and, at a given moment, it drops us, even if the dream is in the author in the way the text is assumed to be. What we call texts escape us as the dream escapes us on waking, or the dream evades us in dreams. We follow it, things go at top speed, and we are constantly—what a giddy and delicious sensation!—surprised. In the dream as in the text, we go from one amazement to another. I imagine many texts are written completely differently, but I am only interested in the texts that escape. (1993, 98)

Start again.

Hypertext is the confirmation of the visual kinetic of rereading. This is not a good first definition of the form or art but rather one made possible by a kind of prospective rereading that, given a world in which ketchup bottles have websites listed with their ingredients, assumes the reader has at least a muddled sense of hypertext from the world wide web. Hypertext is a representation of the text that escapes and surprises by turns.

The traditional definitions of hypertext begin with nonlinearity, which however is not a good place to start given the overwhelming force of our mortality in the face of our metaphors. Either our lives seem a line in which our reading has ever circled, or our lives seem to circle on themselves and our reading sustains us in its directness and comforts us in its linearity. My own amended definition (1995) of hypertext acknowledged the mortality and turned the metaphor to drama while unfortunately adding an element of the metaphysical: "hypertext is reading and writing in an order you choose where your choices change the nature of what you read" (13).

Our choices change the nature of what we read. Rereading in any medium is a conscious set of such choices, a sloughing off of one nature for another. The computer is always reread, an unseen beam of light behind the electronic screen replacing itself with itself at thirty cycles a second. Print stays itself—I have said repeatedly—electronic text replaces itself. What hypertext does is to confirm this replacement, whether in the most trivial sense in which we as readers sustain the text before us by merely forgoing the jittery shift of mouse button or Page-

Down key, or in the deeper sense, itself shared with rereading in any medium, where we linger or shift back intentionally upon a text, making each reoccurrence or traversal its own new or renewed text, the exploration of a dark seam of meaning that mere choice seems to illuminate and (we hesitate to suggest) create for us.

Each iteration "breathes life into a narrative of possibilities," as Jane Yellowlees Douglas says of hypertext fiction, so that in the "third or fourth encounter with the same place, the immediate encounter remains the same as the first, [but] what changes is [our] understanding" (1991, 118).

Start again.

The workshop query with which I began this essay seeks to isolate a set of primitive choices that both prompt the visual kinetic of rereading in hypertext and at the same time isolate the elements of what Douglas calls a narrative of possibilities. The attempt is to move from the nonce upon some times, not so much telling an old story with new twists, as twisting story into something new in the kinetic alternation of *ricorso,* flashback, renewal. The great advantage of this exercise is that it immediately confronts writers who are often quite skeptical about hypertext fiction with literary and artistic questions about linking rather than technical ones about software. It engages working writers with aesthetic and readerly questions about linking rather than encouraging a choose-your-own-adventure sort of drearily branching fiction.

What I do is to ask the writers to write four parts of something, keeping the notion of parts and something intentionally fuzzy but making it clear we are talking narrative. I ask them to use the hypertext system (in this case, Storyspace, created by Jay Bolter and myself with John Smith) to create four spaces (boxes) for the four parts. I encourage them to do this very quickly and not to worry about how extensive or finished the writing is.

Once this is done, I first have them re-create linearity, that is, link the four parts, not merely to teach the simple hypertextual skill but also to reinforce that in hypertext even the linear is a choice. Then I ask the question with which I began this essay:

Suppose at this point your reader, before going on, has to reread one part of what comes before, which would it be?

No one asks why. There are reasons.

Not the least of which is that writers in my experience contemplate a reader in motion across the space of a text like someone inhabiting a map not as a map but as the rereading of a map that we enact and test in motion. That is, writers imagine readers reading as they read when they reread and rewrite. To try to see this let us consider a simple story, in fact the example I use in presenting this exercise to writers, a sweet, old, and endlessly compelling story in which each part is a single sentence: Two people meet. They fall in love. They quarrel and part. They reconcile.

Suppose at this point the reader of this story, before going on, has to reread one part of what comes before, which would it be?

A writer may decide that having read this story and reached its reconciliation, her reader should reread the second section in which the two characters fall in love. Obviously a variant of this strategy (not necessarily requiring that the exact text be reread) is of course what constitutes flashback. With Storyspace this link involves a visual stitch, in the case of this example a line between the fourth and second boxes on the screen. For a later reader this stitch will offer a way back into the sequence of the text and beyond.

Once the writer has linked back into the sequence at whatever point, she is confronted with the following analytical situation: We can agree, I suggest, that we always have at least a theoretical fifth space in mind at the point where we intervene in the text to require a rereading. This fifth might have been a virtual closure, an understood (if uninscribed) gesture toward an end or 'The End.' This end space I call the metanode. Or in fact the fifth space may be a "next" step (a genuine fifth part to the four parts) that the creating mind automatically or instinctively generates despite the exercise's requirement that there be only four parts. However it may also be that the very act of rereading and thus reentering the text has suggested another direction for the narrative, something that not only recapitulates the story but somehow begins another one newly discovered there or at least disclosed in the repetition. "To come back to the only thing that is different is what is seen when it seems to be being seen," Gertrude Stein suggests (1990, 514), "in other words, composition and time-sense."

Once there is a general understanding of these possibilities, I am also prepared to suggest that for the writer only three possible kinds of links exist from the place where we have linked back into the story (in my example only three possible outcomes from our retrospective look at reconciled lovers first falling in love).

If after rereading we go from 2 to 3 again (or in fact any part of the four-part sequence), the link is a *recursus* (or cycle), often a stratagem of modernist/absurdist fictions. This ricorso (the nonce: to the begin again) takes the modernist turn around the track, where the mind loops, by commodius vicus of recirculation ever across the space of the same text, with an implicit promise that there is more to be seen in the turning and that we are not (or are, it is the same thing) looped like Yeats in the loops of brown hair.

If we leap from 2 to the next node, across 3 and 4 to the uninscribed fifth space held in mind, then the link is a flashback. That is the story resumes its intended course (or ends) refreshed by this new look at previous thematic material. Flashback (the next: a leap to the metanode, onward or ending) is an old friend, alternately refreshing or confirming our sense and indeed the experience of a previously viewed episode. It is the woven etymon, text as textus.

If we go from 2 to the new space, not an imagined fifth but escaping inward and outward simultaneously, then the link is a renewal. Linking itself—rereading itself—has discovered and opened a story dimension. Renewal is not textus but narrative origami, where what opens and renews is not the inscription but the narrative of possible inscriptions. This space in which the visual kinetic of rereading unfolds is one that the computer offers a medium for which it is uniquely, though not exclusively, suited.

To be frank the workshop is always both a fascinating and a mildly disappointing exercise, a bit of formalist conjuring in which all the ballet of the three-card monte is lost in the mere shine of the face cards. That there are three link primitives does not speak to their myriad types of course. Of recursus, there is hallucination, déjà vu, compulsion, riff, ripple, canon, isobar, day-dream, theme and variation, to name a few. Of renewal there is the death of Mrs. Ramsey and the near disintegration of a house, the chastened resumption of the Good Soldier, Leopold Bloom on a walk, and a man who wants to say he may have seen his son die. Of the renewal there is every story not listed previously, the unrecollected whisper of your mother, and the barely discerned talk of lovers overheard at the next table as they eat potstickers and drink bad Chinese beer.

The real task of the workshop is thus for the writer to reread the inner folds of sequence and possibility and to fashion what follows from her decision to reopen the text, especially if she has not decided upon the cycle, but rather the metanode or even more compellingly the renewal.

> Not "Revelation"—'tis—that waits,
> But our unfurnished eyes—
>
> —Emily Dickinson

At this point most writers see that, once the text has been revisited and either the new space or the delayed closure of the metanode has been created, the second space now cries for some way to shape its reading for different readers. We want the reader newly come into this simple story to proceed briskly through its inevitable narrative, pause at the reentry, and then leap, without orbiting endlessly unless that is our intention. In any text there are ways to do this, by inference, suggestion, rule, music, or seduction. To these hypertext adds memory and resistance. Story-space and other complex hypertext systems let a writer set conditions that shape the reading according to simple rules that match the reader's experience of the text against the possibilities it opens to her. In a richly linked hypertext these rules (in Storyspace they are called "guard fields") can compound. While a local reading may be as severely shaped as a sestina or a fugue, the permeability of the hypertext makes even a rigorous sequence contingent. You can link in and from any point. A reader may have sailed to this first star from another constellation for which this one forms the third part of a cluster so thick it seems itself a single star and this first star of ours a mere bright spot on its surface. A hypertext fiction spawns galaxies where such constellations link and spin, where other lovers meet and quarrel and part or live forever according to other local rules. This whole dance of complication finally folds in on itself, not in a black hole but a shower of possibilities.

The leap to the new introduces the paradox of hypertextual rereading. Hypertext fiction in some fundamental sense depends upon rereading (or the impossibility of ever truly doing so) for its effects. Yet in a sufficiently complex and richly contingent hypertext it is impossible to reread even a substantial portion of the possible sequences. Indeed for any but a reader who has consciously blazed her way through the thicket (breadcrumbs, in fact, have become a technical term for computer tools designed to keep track of the reading of hypertexts) it is unlikely that successive readings by a single reader will be in any significant way alike. Even in less vigorous hypertext systems such as current instantiations of the world wide web, bereft of the systematic memory that shapes possible readings, the linked surfaces of possibility themselves compound. Despite the most earnest efforts of so-called human factors specialists, and despite the earnest accumulation of lists, breadcrumbs and bookmarks, and other virtual aide-mémoire within the interfaces of web browsers, the narrative of possibilities unfolds. Even the flattest list of visited web pages is thick with possibilities and mixed sequences, as suits and melds are folded within a deck of cards dealt upon a table.

The reader's task in hypertexts becomes a constant rereading of intentions against the rereading of elusive or irrecoverable sequences. We see and lose our hopes for the text by turns in the shifting screens. Again this experience is not exclusive to electronic texts but rather one for which the computer is uniquely suited and within which the inevitable exchange between our intentions and our recognition of the text's possibilities becomes more transparent. "Genuine books are always like that," says Cixous, "the site, the bed, the hope of another book. The whole time you were expecting to read the book, you were reading another book. The book in place of the book" (1993, 100).

That which is reread is that which is not read. To read the

book in place of the book is not to read the book placed (by whom?) in the scope of our expectations. As is her practice Cixous seamlessly moves from reading to writing, seeing in the exchange between them a recognition of mortality, which is to say the body. "What is the book written while you are preparing to write a book? There is no appointment with writing other than the one we go to wondering what we're doing here and where we're going. Meanwhile, our whole life passes through us and suddenly we're outside" (1993, 100). It is we who place ourselves retrospectively within the scope of our expectations. Retrospective expectation is fundamental to the experience of rereading in any medium. Outside, our lives passed through us, we are nostalgic for a complex tense in which what was can be again what will have been other than what it is. Like hypertext, the tense disappears in the parsing, we both cannot and must reread the whatwas that will have been otherwise.

Yet it isn't difficult to do the impossible. We relive our lives in reverie aware that we cannot embody dreams. We reread any text in humility, not only aware that we cannot recapitulate our original experience of it but also that the experience itself was originally unsubstantiated, its evidences lost not merely to history and memory but to even surface recognition. *Où sont les mêmes d'hier?*

Start again.

It isn't difficult to do the impossible. It seems merely literary stratagem, the artifice of the avant-garde, to claim that the experience of a new textuality is somehow not reproducible in the old. Innovation (whether literary or rhetorical) reads and is read by what it extends, alters, ignores, or supplants. With enough rereadings it isn't clear that anything has really changed. In Milan some years ago an exhibit of Rodchenko seemed staid and even conventional to eyes used to computer graphics and morphing fonts; the Constructivist project seemed a matter of the organic quality of pre-offset inks, the geometry of hand-ruled typographical elements, the counterplay of red and green inks against yellowing papers. We reread the prospect of change from the vantage of change and find it wanting, its fulfillment robbing it of its possibility.

Yet this is not simply an aside about the place of hypertext as a literary experiment, but rather an attempt to isolate a distinctive quality of the experience of rereading in hypertext. The claim that hypertext fiction depends upon rereading (or the impossibility of ever truly doing so) for its effects is likewise a claim that the experience of this new textuality is somehow not reproducible in the old. The question at hand is not whether print textuality anticipated or can accommodate innovations of electronic textuality (it did and often can) but rather whether distinctive differences in reading, and thus rereading, characterize each of them.

To see differences, however, it is not necessary, or even helpful, to argue for or against succession (this is why an avant-garde always dissipates: it means to become what it wishes to end). Instead it may be useful, and surely is symptomatic of our age, to argue for parallelism and multiplicity. Differences show as differences are allowed. As Mireille Rosello notes, "the delay in the emergence of new knowledge may also be the condition of its future growth. Rather than imagining our period of transition to hypertext as a point where something old is replacing something new, I would be content to see it described as a time when two ways of reading and writing, and two ways of using maps, are plausible at the same time" (1994, 151).

This is to see change as something different (the self-referentiality here intended), as if Rodchenko created computer graphics at the same time that he and other Constructivists pushed the limits of print. (Richard Lanham [1993] argues as much for the Futurists.) Or for the matter at hand, this is to say that an independent system of reading exists in parallel with the current system of reading in hypertext; that they do not so much confuse each other as enhance each other; and that they do not promise the extinction of one or the co-optation of the other but rather the permeation of each other. More importantly (or closer at hand) the system of reading hypertextually is intimately related to what is called rereading in the parallel system of reading print.

There is of course another argument I want to make here, or have been making though not overtly, and it is that reading in hypertext means to re-create the writers experience of rereading in the process of composing printed works. In fact many hypertext rhetoricians, critics, and theorists assert this claim with greater or lesser elegance and subtlety. Many commentators append the initial of the writer in the inelegant formulation of "wreader" to characterize the new system and its roots. My own well-known notions of exploratory and constructive hypertext are only slightly more subtle. In the sentence before those quoted above Rosello makes a distinction between screening and reading texts: "For a long time, I suspect, the activity of reading hypertexts (rather than screening them) will be considered acceptable and normal."

In her term *screening* Rosello wants to recover something like the seamless move between reading and writing that I have suggested Cixous sees in our embodied mortality. It likewise means to evoke (and in fact is probably the source of my sense of) how the reader in motion across the space of a text inhabits a map not as a map but as the rereading of a map that we enact in (and as) our bodies. Yet there is a trap in the seam in the seamless, and the map infects the body that enacts it. Rosello speculates whether "a new geometry of space is needed in order to invent communities that will have little to do with proximity and context" and it is a speculation that spills, as light through a screen, on the image of the writer rereading seen as the reader writing. We are not the writer because we are reading, we are not the reader because we are writing; the questions at hand are ones of proximity and context. "While the noun *screen* connotes an outer, visible layer, the verb *to screen* means to hide," writes Alice Fulton:

> The opposing definitions of screen remind me of stellar pairs, binary stars in close proximity to one another, orbiting about a common center of mass. Astronomers have noticed a feature common to all binaries: the closer the two members lie to one another, the more rapidly they swing about in their orbit. So screen oscillates under consideration. (1996, 111)

Start again a last time (at last?).

I know when I confuse, at least when I confuse myself. I reread Rosello and Fulton desperately seeking the thread I saw there, not in them but in an argument as yet unwritten. I reread them for my intentions but then worry that I have misread their intentions. I intend to read them and yet leave unread what I mean to see there. Knowing my confusion is often as much as I can hope for in my rereadings.

In the confusion of reader and writer there inevitably lies the confusion of characters in a fiction, the confusion of episodes in its sequences, the confusion of voices in what we attribute to ourselves as a dialogue. It is not a literary stratagem but a matter of fact that the particular experience of the new, albeit parallel, textuality of reading hypertexts is somehow not reproducible in the old. I said that differently, in inverse, before above. You can reread and find out as much, or perhaps had kept it in mind long enough to notice as soon as it occurred.

This is not entirely possible in hypertext. You can neither always go back above, or in fact count upon the existence of the same "above" from reading to reading. What follows from this, of course, is that you cannot always count upon the applicability of what you keep in mind to what follows upon the choices you make based upon that mindfulness. Mary Kim Arnold's hypertext fiction "Lust" is not much longer than a poem (although it starts with one, it is not one), something short of eighteen hundred words in thirty-seven screens (or spaces), fugal, multiple, confusing (even for some readers, my students especially, maddening), haunting, irreproducible here although I could easily (with permission) include all its episodes.

In the story a woman has hurt a man or the man her, there is a knife and blood and gravel and a rug, they have a child or she thinks him one or he does her or they each or both imagine or desire or recall when they were one, he abuses her or she him or we imagine as much, they make love or do not, they sleep or do not, she runs off or he brings her back, they may or may not drink orange juice. There are men named Dave, John, Jeffrey, and Michael; the woman is unnamed, always called "she." It is possible to read all thirty-seven screens in a single reading and possible to read for a very long time without seeing one or a substantial number of the screens. Sometimes the same screens appear in the same order but interrupted by different sequences between them. There are 141 links among the thirty-eight spaces, thirty-six of which begin with the individual words of the following poem, which is the entire text of the first screen:

Nearly naked
this summer night
sweet and heavy,
he comes to her.

This night, she follows him,
sweat between them.

They speak of the child
and the summer sun
with words that yield
to the touch.

('Prologue')

Let us concentrate for a moment on a single, simple, screen entitled 'He and the Child' (note that screen names are given, indicated here by single quotes, since hypertexts are, largely, unpaginated). 'He and the Child' engages us with one of the less controversial, seemingly more easily apprehended plot elements. "He is gentle with the child. Speaking softly, deliberately, muscled arms embracing soft naked flesh." This is the text of the screen in its entirety. It happens that there are five screens that can lead a reader to this one (although to know this you have to radically reread or unread the text, in fact dissect it within the Storyspace program in which it was created). The first leads directly from the word "that" in the second last line of the 'Prologue' poem; a click on that word will change the screen to the text of 'He and the Child.' A reader can also reach this screen coming from a screen called 'He Expects,' but only if the reader has previously read a screen called 'Touching,' which has four links leading to it, including one directly from the word "sweat" in the sixth line of the 'Prologue' poem (not from the word "touch" in the last line, which instead leads to a screen called 'Penis') but which itself does not lead directly to the screen called 'He Expects.'

A third link to the screen 'He and the Child' comes from a screen entitled 'Innocent,' which has also four links into it (one from the prologue) and two other links from it, including one (to a screen called 'In Noce') that is followed if the reader has already encountered the screen called 'He and the Child.'

And so on.

No one reads this way, of course, except the hacker or the literary critic. Or perhaps the writer. Although no one writes this way, to read this way while writing is to reread as a prospective reader and in the process unread the text in favor of what is not normally read within it.

The combined text of one sequence of the screens mentioned above would read (does read) as follows:

> He touches her. He touches the child. The child screams.
> She touches the blade of the knife to him, cold, smooth. He does not speak.
> He screams. The child does not speak. The child picks up the knife. There is no blood.
> There is no child.
> There is only morning.
>
> ('Touching')

> He was nearly naked, except for the baseball cap. He does not speak to her.
>
> He expects her to come to him.
>
> ('He Expects')

No one reads this way, of course. First, in this sequence the text is always bracketed, in the case of the sequence described above by screens 'Prologue' and 'He and the Child' that I have not quoted again in this text since they have already been seen. The sequence is also bracketed by the act of the mouse click or key press and the flicker and shift of screens that confirm the intention of the reader to go on. Pages likewise settle and sigh though we no longer account their confirmation.

Second, no one (or only one in thirty-six readers making the same choice at the prologue) comes upon this sequence in this order unless by chance, while any number of readers can come into this sequence at the point past the prologue (through another path to 'Touching') and some readers can come to 'He and the Child' (or 'Touching') having already seen either of them in another sequence that, unlike the one from 'Innocent,' for instance, makes no account for a reader who has already encountered the screen called 'He and the Child.'

The hypertext sometimes recalls what the reader has read and sometimes not, but obviously only in a systematic (we might almost say mechanical, were it not a silicon-based slab of light) way. That is, if such an inconsistent if not contradictory recalling can be called systematic. Hypertext builds a systematic level of the literal upon the experience of rereading, with words like *recall* and *recollect* taking on (reassuming the name, rereading the sign) their literal meaning: When a text is recalled by the system, the recollection remains within the reader.

"Assembling these patched words in an electronic space, I feel half-blind," Shelley Jackson writes in her hypertext novel, *Patchwork Girl, or, A Modern Monster,* a work attributed to Mary/Shelley and Herself. It is part of a long section—linked contingently and multiply in much the same manner that "Lust" is—called "body of text." Most hypertext fictions include these self-reflexive passages. The section seems to oscillate in its voices among these three attributed authors, and at least once engages in a dialogue with (a text of) Derrida. Despite this, I think these passages are something more than a postmodernist token for the pinball game of blur and blink; and, to the extent that my own work can be seen as wellspring, they do not I think merely mark the fledgling stream (a flow is hardly a tradition) of a passing form in an uncertain medium. Instead, or more accurately concurrently, these passages are also a gesture toward a parallel system of reading that invites the reader to read as the writer does rereading. It is, says Jackson

> as if the entire text is within reach, but because of some myopic condition I am only familiar with from dreams, I can see only that part most immediately before me, and have no sense of how that part relates to the rest. When I open a book I know where I am, which is restful. My reading is spatial and even volumetric. I tell myself, I am a third of the way down through a rectangular solid, I am a quarter of the way down the page, I am here on the page, here on this line, here, here, here. But where am I now? I am in a here and a present moment that has no history and no expectations for the future. (1995, 'This Writing')

No expectations except motion, sequences bracketed by the act of mouse click or key press, the kinetic of rereading. "Or rather," Jackson continues, "history is only a haphazard hopscotch through other present moments. How I got from one to the other is unclear. Though I could list my past moments, they would remain discrete (and recombinant in potential if not in fact), hence without shape, without end, without story. Or with as many stories as I care to put together" ('This Writing'). If no one reads hypertext by dissection (although Jackson's story from time to time literally dissects both Mary Shelley's monster and Frank Baum's girl cut and repatched), how does the reader mark this hopscotch of history, the kinetic text? In a voice that anticipates (or participates in the same swirl that engenders) Alice Fulton's, and that likewise recalls (or recollects?) my own suggestion above of the renewal link as narrative origami, and that finally marks the commonplace poetry and virtuality of the sewing pattern, Jackson's tripartite narrator suggests that we read along the dotted line:

> The dotted line is the best line:
> It indicates a difference without cleaving apart for good what it distinguishes.
> It is a permeable membrane: some substance necessary to both can pass from one side to the other.
> It is a potential line, an indication of the way out of two dimensions (fold along dotted line). In three dimensions what is separate can be brought together without ripping apart what is already joined, the two sides of a page flow moebiusly into one another. Pages become tunnels or towers, hats or airplanes, cranes, frogs, balloons, or nested boxes.
> Because it is a potential line, it folds/unfolds the imagination in one move. It suggests action (fold here), a chance at change; it also acknowledges the viewer's freedom to do nothing but imagine. (1995, 'Dotted Line')

What we read is suggested action, one way out of the two dimensions, a gesture toward Rosello's "new geometry of space" beyond proximity and context. It is this gesture that Jackson marks in the only link from the space 'Dotted Line,' a link that discloses the dots in its lack of gap, the directness of its bracketed action for the reader: "I hop from stone to stone and an electronic river washes out my scent in the intervals. I am a discontinuous trace, a dotted line" ('Hop').

Poet, hypertext theorist, and computer scientist Jim Rosenberg in his poetic sequences *Intergrams* (1993) and *The Barrier Frames* (1997) has created new poetic textuality that is quite literally not reproducible in this older one. His poems flicker and focus from a dark sea of blurred and overprinted language as the mouse moves over their surfaces, clearing suddenly into discernible patches, like the backs of golden carp rising briefly to sunlight in a dark pool or floating into focus like the fortune cookie scraps of text of the old prognostic Eightballs. No sooner do they snap into clarity than with the least movement they are lost again and again as soon as they are gained. His is, he says, a hypertext of "relations rather than links" (1996a, 22) and it is no wonder that, when he comes to propose a hypertext poetic, it is one that attends to action. His paper "The Structure of Hypertext Activity" argues that "readers discover structure through activities provided by the hypertext" and offers a three-level taxonomy of the activities of reading from acteme to episode to session, where his coinage is that "acteme is an extremely low-level unit of activity, like following a link" (1996a, 22).

Or rereading, which in hypertext rhetoric becomes dissected (along dotted lines) into varieties of "backtracking": "One may revisit a lexia simply to read it again," says Rosenberg, simply throwing the baby of this current essay out with the golden carp's dark bathwater, "or it may be a genuine 'undo,' perhaps the reader didn't mean to follow that link at all." His immediate, low-level interest here is in how to represent the meaningfulness of an action for the reader. "These [backtrackings] are arguably different actemes," he says, "though typically not distinguished by the hypertext user interface" (1996a, 22).

His higher-level interests, however (or are they the Chomskyan deeper structures? or perhaps instead—simultaneously—the topsy-turvy, each-side-up, permeable membrane of body or screen?), are in episodes and sessions where "the episode itself *emerges* from reading activity" (1996a, 26) and where, in place of closure, the reader may at the end of a session "obtain a sense of completion about the gatherings, i.e., the reader's sense of completion is exactly a writer's sense of completion: the gathered result 'works' artistically as is, now is a good time to stop" (28).

Under such conditions rereading and unreading are alive in contention, and vie like subjects of a fugue. "Whether an instance of backtracking is really an 'undo' may be rephrased," says Rosenberg:

> Does backtracking *revoke* membership of actemes in an episode? It depends on the circumstances both of the hypertext and the reader's frame of mind. The reader might revisit a previous lexia to read it again—perhaps for sheerly "musical" repetition, or to reread a prior lexia based on some resonance or reference in the present lexia. Here one might argue that all the backtracking history is part of the episode. Or, the reader may be backtracking to undo having arrived at the current lexia by mistake—backtracking to *remove* from the episode the acteme that caused arrival at the current lexia. The episode is thus a combination of history through the hypertext, the reader's intention, and the reader's impression of what "hangs together." (1996a, 24)

Emerging meaning gathers in episodes that combine in sessions of reading and rereading and sometimes seem (to the reader) to mean on their own. "There is a kind of thinking without thinkers," says Jackson. "Matter thinks. Language thinks. When we have business with language, we are possessed by its dreams and demons, we grow intimate with monsters. We become hybrids, chimeras, centaurs ourselves" (1995, 'It Thinks'). Such a thinking without thinkers occasions Rosello's notion of screening as well. She speculates about "what kind of context is being created as the result of experimenting with apparently arbitrary connections . . . not . . . the kind of arbitrariness that comes from conventional forms and discourses, but rather a deliberate incursion into the messy realms of chance, random connections and meaninglessness" (1994, 134).

Traditional definitions of hypertext begin with nonlinearity, which however is not a good place to end given the overwhelming force of our mortality in the face of our metaphors. "I align myself as I read with the flow of blood," says Shelley Jackson's triple narrator,

> that as it cycles keeps moist and living what without it stiffens into a fibrous cell. What happens to the cells I don't visit? I think maybe they harden over time without the blood visitation, enclosures of wrought letters fused together with rust, iron cages like ancient elevators with no functioning parts. Whereas the read words are lubricated and mobile, rub familiarly against one another in the buttery medium of my regard, rearrange themselves in my peripheral vision to suggest alternatives. If I should linger in a spot, the blood pools; an appealing heaviness comes over my limbs and oxygen-rich malleability my thoughts. The letters come alive like tiny antelopes and run in packs and patterns; the furniture softens and molds itself to me.
>
> (I do not know what metaphor to stick to; I am a mixed metaphor myself, consistency is one thing you cannot really expect of me.)
>
> What I leave alone is skeletal and dry. (1995, 'Blood')

Screeners and gatherers, we do not know which metaphors to stick to, although the body is our type for stick-to-itiveness. And so, autumnal readers we wait for the leaves to fall. What we leave, alone, is skeletal.

Start again (backwards, back words).
"We live," says Shelley Jackson,

> in the expectation of traditional narrative progression; we read the first chapters and begin to figure out whether our lives are romantic comedy or high tragedy, a mystery or an adventure story; we have certain hopes for our heroine, whose good looks can be expected to generate convoluted formations among the supporting characters and indicate the probable nature of her happy ending; with great effort we can perhaps lean sideways and veer into a different section of the library, but most of us do our best to adhere to the conventions of the genre and a kind of vertigo besets us when we witness plot developments that had no foreshadowing in the previous chapters; we protest bad writing. (We are nearly all of us bad or disorderly writers; despite

ourselves we are redundant, looped, entangled; our transitions are awkward, our conclusions unsubstantiated.) (1995, 'Lives')

In the process of reading for what she has not written (or written well) she often does not read what she has written well (or not written). Most often she looks for the thicket, the paragraph or phrase that relinks a vision or reforms it, a vision that she put aside or lost, which dwindled or lapsed, which exhausted her or she exhausted. The writer rereads and unreads in the same scan, sometimes looking for the place that needs attention, other times seeking surprising instances of unnoticed eloquence that her attention now confirms in a process of authorship. That which is reread is that which is not read.

Chapter 8

On Boundfulness: The Space of Hypertext Bodies

The Chapter I Am Writing: Heterotopic Dwelling, or, I Am Here Aren't I

> The things he showed were primarily traces; the photograph alluded to knowledge that it would scarcely show. It took the visible tip, a detail or a surface, and found other ways to indicate what was not there, what had been relegated to the distance, when not altogether cut. This was the thrust of emptiness in the third city.
> —Molly Nesbitt, "In the Absence of the Parisienne . . ."

Some time ago I was sent the text of the proposal for a collection of essays that included, among other chapter abstracts, the following:

Ethereal Texts: words in the web (*) Michael Joyce, Vassar College, or Mike Crang, University of Durham

The possibilities of virtual geographies are not simply "out there" to be commented upon, they also inform the nature and possibilities of those commentaries. Developing issues raised in the previous chapter, this essay will explore the ways in which recent developments in electronic media might be affecting the ways in which these virtual geographies are represented. Looking specifically at the world of hypertext, the chapter takes up the argument that hypertext offers a microcosm of the web as a whole—that is, that it takes the form of links between disparate fields of knowledge in an electronic space. It asks how hypertext shapes these links, about the new constellations of knowledge that its geography makes possible and, more tentatively, about the power to the reader it makes available to radically reorder the text to create new, perhaps more open forms of knowledge. But it sets its enquiries in a historical perspective that questions how far such developments do indeed move beyond older forms of text and modes of textual representation and reading.

This is, of course, the chapter I am writing (and now that it is collected a second time can be said to have written), although at the time it was proposed to me it could have been someone else who was writing it (the someone who had written the abstract, one supposes, perhaps the editor of the collection where this essay originally appeared, Mike Crang, who is named here, although someone else could have likewise given its injunction to him). A year or so ago I received a similar solicitation from an editor to contribute a chapter regarding hypertext for a proposed collection on rereading. That solicitation was addressed to Michael Moulthrop, a conflation of my name and that of the hyperfiction writer and theorist Stuart Moulthrop.

It has ever been thus for editors and authors and proposals and collections and so I don't suppose to suggest that these anecdotes necessarily limn "the possibilities of virtual geographies . . . [that] inform the nature and possibilities of those commentaries."

Yet the plausibility of interchangeable authorship is also the plausibility of interchangeable identity.

Well, I'm here, aren't I?

The strategy of making myself the center of a text about boundfulness is suspect in this late age of premature post-postmodernism. However, we may ask to whom the identity of this "I" matters (and likewise who is this "we" that may ask when she's at home). As Habermas says, "Even collective identities dance back and forth in the flux of interpretations" (1992, 359). A chapter on hypertextuality is wanted and whomever (preferably someone with at least one name beginning with *M*) may write it.

A chapter on hypertextuality is found wanting. Hesiod saw the world as founded (or found the world he saw) upon wanting.

At the level of style, it may be that a certain readership exists who finds this particular kind of lyrically self-reflexive playfulness identifies a certain "author," namely me. Without doubt, I can number myself among such readers. The number is either one or the zeroeth.

At the level of genre this one is a variety of "My-story," Gregory Ulmer's (1989) intentionally artificial, protohypertextual genre of successively (and heuristically) generated spaces.

At the level of general organization (we could say argument or organic unity were we more old-fashioned; we could say apparent organization or emergent form were we more postmodern), any of us would do. Although it must be supposed that I have done as well or better than at least one other in this instance since the editor, one such other, has seemingly allowed this entry (or so I surmise, however projectively, as I write this).

I can find some consolation in set theory I suppose, since for each of the two collections, I am the one recurring element in the set of possible writers. But in fact my putative authorship (insofar as it is not included in the otherwise seemingly plentitudinous authorships that open out from the questions at hand) merely marks the bounds of the questions that are examined here, something Mike Crang's abstract did long before me and that in some sense I therefore always fulfill if only in the suggestion of its absence, a presence that my authorship metonymically enacts.

The reader functions likewise, here as well as in hypertext, although there is less chance that the text here will slip from its putative authorship within the confines of the chapter.

Ethereal Texts: words in the web, by Michael Moulthrop or Hesiod.

I'm here, aren't I, at least until page _____ (the last word one I am incapable of writing since the page itself does not exist at the time of this writing and so therefore is tokenized, marked by the dash as a sign of some future authorship, which may or may not be filled).

The author serves as the singular. *The* space of hypertextual bodies in the platted (plaited, plaintive) provocation of the singular title of the chapter "On Boundfulness: The Space of Hypertext Bodies" within which this chapter "Ethereal Texts: words in the web" does and does not appear.

The section you are reading, or have read, depending on how you constitute this coda, originally appeared at the end of the chapter I am writing, until the original editors intervened with the congenial suggestion that "if you moved the last section about your authorship to the front it might prepare the reader about how to interpret the essay . . . [since] unless they read the editorial (pious hopes) they won't know in advance this is about hypertext."

This is about hypertext.

Thinking One Self Else Where

Seasoned and bendy
it convinces the hand

that what you have you hold
to play with and pose with
   —Seamus Heaney, "A Hazel Stick for Catherine Ann"

This particular Irish literary genre, the immram, deals
with high-spirited ventures into an unknown ocean. . . .
However fantastic the adventures and imaginary the
geography, the immram record experiences that
were real.
   —Carl O. Sauer, "Irish Seafaring"

Dieberger doesn't think he's here—what could this mean?—he thinks he's somewhere else. It doesn't do to ask him, I can tell you. He's not alone in thinking this, wherever he is. Let's ask him.

He wants to be everywhere, he wants to be nowhere. Utopia = pantopia. He has no leg to stand on. He's ectopic, "beside himself" as the saying used to go. He's used to it, he's a juggler.

He's the new wave, it says so on the program. Likewise, likable, he waves to the crowd as he, slightly Teutonic (i.e., "I dun't read anysing . . .)—Austrian actually, says:

"I don't read anything on the web anymore. I just check out the links and mark the ones I want to come back to later. Though I never do really."

The betweenness that is not strictly nomadic nor instrumental nor algebraic but rather volitional, constitutive. The break of a link in a web conceived as everywhere impinging surfaces, a topological space, skin, surface of surfaces without surface, amen.

It is important to ask ourselves how we constitute the body

that sees itself in boundfulness, permeable though not less permanent than the rest of the impermanent world, wholly interstitial and yet no less whole.

Ectopic≠osmotic.

Not by negation. It is something other than the nomadic body that, however deterritorialized, occupies its own space.

Marcos Novak thinks not:

> I coin the word pantopicon, pan+topos, to describe the condition of being in all places at one time, as opposed to seeing all places from one place. The pantopicon can only be achieved through disembodiment, and so, though it too speaks of being, it is being via dis-integration, via subatomization of the consciousness, rather than by concentration or condensation. (1996, n.p.)

It is something more than a Berkleyean turn (the bishop not the campus) to say that Dieberger (the mountains) means differently. He's no places at many times and though nowhere all at once elsewhere here.

Hear hear.

His pronouncement is met with great applause (a great many boys in the audience claim via huzzahs that they too no longer read links, just swing from url to url like Tarzan; meanwhile back at the main camp Jayne Loader, a WebWench but no Jane, builds a literature of links upon the boyish schema—not Schama, no Simon upon which is built a church—of proto-(pre-frame)-Suck.

There are certain kinds of literacies that can unpack the previous sentence, though they are not coterminal merely coincidental. (Tarzan and Jane are, of course, the pre-post-colonial icons of certain pre-post-feminist baby boomers; Jayne Loader's Web-Wench site [http://www.publicshelter.com/wench/] exploits a contrapuntal aesthetic first popularized by the webzine *Suck*, namely www.suck.com, in which imbedded links smartly play off a smart-ass text, the pleasure being in the simultaneous recognition of the implicit disjuncture and plenitude of the web. [Modernity being, as Edward Soja puts it, both "context and conjuncture" (1993, 147)], of which disjuncture is but a special case; though when I first read the phrase I confess I thought he said conjecture.) Loader's film *Atomic Cafe* surely is among prewebbed predecessors of this aesthetic of interwoven disjuncture and plenitude—Novak calls this shift of consciousness "from the society of the centripetal panopticon to the society of the centrifugal pantopicon" a Centrifug(u)e.

The apposition of the initially innocent sentence and the long, dense, and overdetermined parenthetical paragraph is not merely a stratagem of modernism, duly (and dully) carried forth through postmodernism and hypertextuality but also itself an instance of boundfulness.

To the extent the pantoptic young men on the flying trapezoids move from link to link without alighting there may, of course, be reasons to suppose that the centrifugue is the same old song. What, we could ask, is the meaning of the swan song of the one who is moving on? What production, mea culpa mea *marxisme,* is involved in passing over, links or space? In a proto-post-Marxist melody of his own, David Harvey suggests that "denigration of others' places provides a way to assert the viability and incipient power of one's own place." Young anthropoids make such calls. All this frantic swinging from branch to branch may merely be place-marking in what Harvey calls "the fierce contest over images and counter-images of place" and where "the cultural politics of places, the political economy of their development, and the accumulation of a sense of social power in place frequently fuse in indistinguishable ways" (1993, 23). In a consideration of hypertext browsers (here meaning programs like Netscape or Internet Explorer) in which the reader "cannot change the text but can only navigate among already-configured trajectories" as instances of de Certeau's wanderers *(Wandersmänner),* Mireille Rosello likewise finds an underlying production in which browsers (that is, readers) "passively consume what others have produced or written [and] the steps of the walker across the city are disembodied, like weightless information saturating a network" (1994, 136). She evokes de Certeau's assertion, "Surveys of routes miss what was: the act itself of passing by" (1983, 97), suggesting, "Like a Derridean trace, such maps keep the memory of an absence. . . . The trace left behind is substituted for the practice" (1994, 136). For the pantoptic young men, that trace is embodied, is the body in its boundfulness. They are walking (or flying) maps of unexplored meaning.

The gesture of the appositional otherhandedness and scholarly stitchery, on the other hand, closes off a dialectical ticking that bounds the feathery space upon which lie kinds of systematic literacies (no less [post]modernist) that can pack and unpack the previous sentences and paragraphs like the suitcases of some wanderer, a foreigner for instance, making repeated journeys elsewhere without settling and, outside our seeing, disappearing without a trace.

I've elsewhere suggested as a figure for hypertext Kristeva's characterization of the "otherness of the foreigner" in terms of "the harmonious repetition of the difference it implies and spreads" in a fashion that she links to Bach's toccatas and fugues, "an acknowledged and harrowing otherness . . . brought up, relieved, disseminated, inscribed in an original play being developed, without goal, without boundary, without end. An otherness barely touched upon and that already moves away" (1991, 14).

Though the otherness may be seen as without boundary, the foreigner is the image of boundfulness as well. Persistence of vision, for instance, or simply simple repetition.

Foreigner or fugue, especially as these are repeated, make space that is both within and somehow simultaneously outside the space of the text. The gesture of the parenthetical, the dialectic, the thematic, the rhythmic, the fugal, the isobaric, the metonymic, the list, the link, the litany, as well as any and all other—whether em-dashed or no—appositional stitchery constitutes the space of hypertextuality. Boundfulness, in this sense, is space that ever makes itself, slice by slice, section by section, contour by contour; never getting anywhere is Dieberger on the links (his hole-in-one a torus, ever a single surface).

[Internote: in the course of looking up Zeno of Elea in order to make the link between this space of ever opening boundfulness and his famous paradox, I discover that when text is copied from the Microsoft Bookshelf 1996–97 hypertext version of *The Concise Columbia Encyclopedia* into a document, at least within Microsoft products, the program automatically (and invisibly in most views) generates a footnote containing copyright information. It is of course possible to imagine that someone copying terms from the Microsoft Bookshelf into the span of terms previously copied from that source could create a fully boundful text of successive citational edges, not unlike topographic contours. Quod erat demonstrandum.]

Dieberger is Andreas Dieberger, whose screen name (and avocation) is juggler, author of "Browsing the WWW by Interacting with a Technical Virtual Environment" (1996). Here (or there) is the ACM Hypertext '96 panel on "Future (Hyper)Spaces."

Dieberger filled in at the last moment for Mark Pesce, cocreator of VRML, virtual reality modeling language (a language as yet apparently unable to put him in two places at once), and so Dieberger is not listed as a participant in the program. His comments are not printed or posted on the web. I may have made them up. You had to be there.

The Phantom Limb and Metonymic Imagination

It is not a matter of deciding to go into cyberspace.
We are always already in it, before the literal condition.
   —Stephen Perrella

I just dropped in to see
What condition my condition was in
   —Kenny Rogers and the First Edition

Even so we could consider the postprior experience of the current condition.

The space of the node—not properly a "screen" or a "page"— a meta-element yet computationally exact, at least to the extent that it is locatable, is thus metonymic. It summons the whole of the web of its relations by means of the disclosure of its failure to participate within that whole, which of course does not exist in any coherent state beyond the suggestion of its absence, a presence that the partiality of the metonymic element enacts.

Whoa. Woe. Woebegone. Wo bin ich?

Let us go through this more simply. A web page currently is represented as a window with some method of indicating depth (scrolling, etc.) that also suggests a measure of local closure or boundedness. Yet that measure (the scroll bar for instance) is not so much metatextual as extratextual. One is not apt to say or think, "I found a picture of a pig seven-ninths of the way down the scroll bar on a fully opened window upon a 640 × 480 monitor." The window itself is merely an artifact of a data structure within the operating system, or the browser, or the html structure (which may itself be a formal hierarchy or a happenchance and opportunistic network as yet unformally represented).

Yet each of these spaces—the proprioceptive measure of a pig seven-ninths of the way down or the unformal abstraction of the containing container—can be said not merely to embody but to constitute the hypertextuality, especially of a random site that for argument's sake I am only presenting through the single discrete element (itself "located" elsewhere, in a gif file for instance, i.e., "This little piggy went to markup . . .").

The experience of this space within the node stands metonymically both for the space of the abstract structures of its representation (window, system, browser, frame) and for the composite space (the site, the web, the story, the reading) within which we experience it.

The insistence upon "click-through" as the measure of advertising effectiveness on the web inverts a rhetoric established by the advertising industry (conventionalized as "the advertising world": the whole earth became a hippie brand name before Stuart Brand became a technological guru) wherein advertising was purported as neutral and extratextual. Thus *Wired* magazine proposes that advertising ought perhaps to enter content (as if product placement remained to be invented). Actually the columnist (I know I should footnote this but, even if this attribution is a lie, the argument, if such, doesn't suffer and the writing in that magazine isn't meant to persist only exist) suggests that advertising be woven through content (as if *textus* meant anything else; as if a weaving were not a penetration; as if a parenthesis were not a lung).

The placement of this particular argument (if argument it be; click-through not yet determined) is a version of the same inversion. Assigning boundaries to this discourse depends upon an invented perception of its meaningfulness. Locality, Arjun Appadurai suggests, appropriating Raymond Williams's term, is "a structure of feeling" (1996, 199). The reader must suppose that the indulgence of the spatialized voice here represents (enters) the content of the argument that is itself represented by the perception of that interwovenness (representation, entrance). Perhaps "I" am simply mad or a sloppy thinker.

Following a talk in Hamburg (53° 33′ N 10° 0′ E) (collected as the first chapter here) for instance a woman comes forward to ask if I am a Buddhist, saying, "I had to listen very closely because I could not be certain from moment to moment whether you were discussing the poem by Milosz, or the scientific concept of catastrophic singularity, or hypertext theory."

"That is the scientific concept of singularity," the *sensei* replied.

The modernist notion that meaning presents itself in (or *après*) fragmentation is ceded (seeded) in postmodernism to a present-tense fragmentation of meaning as a constant stream. Is it fair to ask where this stream is situated? This whole earth? This seven-ninths piglet? the space within a node? the Powerbook in the cinematic twister?

The world of tokenized representations does not exist in any coherent state beyond the suggestion of its absence, a presence that the metonymic element enacts.

The disembodied juggler, all boundful, die Berger, exists in absence. He's on his way not nowhere (the nomos) but elsewhere (osmosis). The endless sectioning of the boundfulness of the web turns Deleuze and Guattari's speculation on the smooth and striated to pure reportage. Osmosis is the liquid equilibrium along a semipermeable barrier.

The web transcends the inevitable spatiality of other hypertexts by becoming primarily osmotic and ephemeral. (Before the hyperfusty bookishness of *Myst,* Rand and Robyn Miller created a transcendently hyperfictional happy-ever-otherworld in *Cosmic Osmo.*) By circling round our senses of confirmation, disclosure, and contiguity, we find ourselves falling into sense. A recognition of traversal prompts my student Samantha Chaitkin to offer "a brand-new metaphor" in a critique of spatial hypertext representations:

"I'd rather . . . jump up into the air and let the ground rearrange itself so that I, falling onto the same spot, find myself somewhere different. Where am *I* going as I read? No, more where is the Text itself going, that I may find myself there" (1996, n.p.).

The "I" is considered constant (as here) and avoids the conventionalized gesture that blurs I/eye and thus either objectifies the world within a gaze or subjectifies the world within a discourse. The ground is given its own authorship when the link is seen as *tour en l'air. En bas,* however, the self stands as metonym on terra firma, part not for but *as* the whole.

The world (ground) (re)assumes its agency in the elsewhereness of our boundfulness (a phrase that in its comic compounding of -nesses summons a lost sense of utopianism, Shaker hymnal phrases, the susurrus of—sweet Jesus—praises). The question at hand is resurrection: what body?

It is difficult not to imagine that something goes on (this ground) elsewhere as we land again and again on a turning world. The most common anxiety that my students report as they come to read hypertext fictions is a feeling that the story goes on elsewhere either despite their choices or unmindful of them.

Hypertext novelist and theorist Stuart Moulthrop early on in the recent history (to make a space) of hypertext literary theory foresaw this metonymic quality of hyperfictional space: "To conceive of a text as a navigable space is not the same thing as seeing it in terms of a single, predetermined course of reading." Moulthrop contrasts "the early intimations of wholeness provided by conventional fiction [that] necessitate and authorize the chain of particulars out of which the telling is constituted" to the distinctive "metaphor of the map" in hypertext fiction that rather than preferring "any one metonymic system . . . enables the

reader to construct a large number of such systems, even when
. . . these constructions have not been foreseen by the text's
designer" (1991, 129).

I've suggested in earlier chapters here that this sort of claim
for hypertext fiction is beyond, if nonetheless beholden to,
Umberto's Eco's literally transformative formulation of the "open
work" that offers the reader an "oriented insertion into something which always remains the world intended by the author"
(1989, 19). Yet even Moulthrop (however necessarily, since at its
earliest stages—a mere decade ago—we all conceived a single
reader and writer in a shifting dance on a single web) argues from
the perspective of the single reader, herself unaware of her com-
peers' actual or imagined progress elsewhere through the shifting
text:

> Metonymy does not simply serve metaphor in hypertextual
> fiction, rather it coexists with metaphor in a complex dialecti-
> cal relationship. The reader discovers pathways through the
> textual labyrinth, and these pathways may constitute coherent
> and closural narrative lines. But each of these traversals from
> metonymy to metaphor is itself contained within the larger
> structure of the hypertext, and cannot itself exhaust that struc-
> ture's possibilities. (1991, 129)

My students' awareness of the simultaneous elsewhere that
they construct from their boundfulness is heightened because
they are aware their choices shape the story, or better still their
continued presence embodies it. "Hypertext navigation," as Terry
Harpold notes, "means not only traversing a space between two
points in the narrative; it means as well electing to diverge from a
predetermined course" (1991, 129).

The phantom limb is the story that goes on elsewhere while
we experience the story of the repeated here. We have the feeling
that we are elsewhere, on another terrain. Yet the elsewhere is
also here; what differentiates metonymy from metaphor is its
suborbital flight, its semantic shift is lateral, a spatializing displacement within the space of the literal without the orbital
escape velocity of metaphor. "The possibility that the reader may
choose to digress from a path of the narrative, and remain within

a field/terrain that is still identifiably that of the text she is reading," says Harpold, "greatly complicates metaphors of intentional
movement that may be applied to the act of reading" (1991, 129).
Simultaneously within and without, the reader doesn't feel her
loss as much as the loss of embodied presence of the body's
boundfulness. Terrain and body alike are systematic longings.

"The body," as Donald Kunze notes, "is not an abstract idea
but a living entity. The body is what the body does, and the body
is thus allied with the process of enactment that fleshes out the
'4th dimension' reinserted into its new position between representation and world." Kunze locates this enactment in "a gap in
the system of Cartesian dimensions . . . between dimensions #2
and #3" where "'spatialized time and temporalized space,' has to
do with the muscularity, enaction, emplotment, and dynamics of
moving from images to solid realities, i.e. the world in which
human action becomes actual, the real and sharable world"
(1995, n.p.).

*Act II, scene iv: The Real enters, accompanied by various lords and
ladies, the Frenchman, and others:*

> There are also probably in every culture, in every civilization,
> real places—places that do exist and that are formed in the very
> founding of our society—which are something like counter-
> sites, a kind of effectively enacted utopia in which the real sites,
> all the other real sites that are found within the culture are
> simultaneously represented, contested, and inverted. Places of
> this kind are outside of all places, even though it might be pos-
> sible to indicate their location in reality. Because these places
> are absolutely different from all the sites that they reflect and
> speak about, I shall call them, by way of contrast to utopias,
> heterotopias. (Foucault 1967, 24)

The speaker is the Frenchman Foucault, and he wrestles with
an angel (although onlookers may perhaps mistake this for a
dance). In his so-called real heteroptopic space I want to situate a
little dwelling, a ritual shed perhaps, in which, for purposes of this
drama, we may distinguish between the flying young men on their
trapezoidal machines and the young woman who spins in the air.

Heterotopic≠ectopic

For Samantha's longing is not, I think, for the body outside the body, the ectopic self, outside temporality and thus beyond mortality. Not a longing for self-evident and embodied truth, a proof. For, as Kunze notes, "by definition, proofs of the body are refused the possibility of detachment; they commit the logical fallacy of self-reference. . . . That is to say, by attempting to step outside the human condition in order to describe any element of it, the mind must falsify its status as body, as member of the subject in question."

Foucault has wrestled with this angel as well, looking for/at the spaces where self-reference can be seen. "I believe that between utopias and these other sites"

> there might be a sort of mixed joint experience, which would be the mirror. I see myself there where I am not, in an unreal virtual space that opens up behind the surface; I am over there, there where I am not, in an unreal virtual space that opens up behind the surface; I am over there, there where I am not, a sort of shadow that gives my own visibility to myself, that enables me to see myself there where I am absent. (1967, 24)

Although perhaps there is nothing to look at (or through). Instead (in the place of) of the ectopic unreality of present absence perhaps what we see is the heteroptopic flicker, the here-there of the woman in the air, the "epistemic shift toward pattern/randomness and away from presence/absence" of what N. Katherine Hayles characterizes as "flickering signifiers" (1993, 73ff.).

Heterotopic≠osmotic

"Interacting with electronic images rather than materially resistant text," says Hayles,

I absorb through my fingers as well as my mind a model of signification in which no simple one-to-one correspondence exists between signifier and signified. I know kinesthetically as well as conceptually that the text can be manipulated in ways that would be impossible if it existed as a material object rather than a visual display. As I work with the text-as-image, I instantiate within my body the habitual patterns of movement that make pattern and randomness more real, more relevant, and more powerful than presence and absence. (1993, 71)

Exeunt the Real, the Frenchman, the spinning girl, et alia.

LEGEND

In an attempt to approximate the space of hypertextuality in this linear form this essay is written in a series of overlays, viz. Lippard 1983, unmarked by typography or, for that matter, any but the most rudimentary narrative or syntactic markers, save perhaps the section conventions of which this is an—anomalous—instance. Thus only the temporal marks this stratigraphy and the claim here, as in any legend—cartographic, folkloric, or mythic, is susceptible to physical observation and verification. Even so I intend a perceptible differentiation of the spaces here, not unlike a mapping, or better still something akin to the "global cultural flows of Appadurai (1996, 33), i.e. (a) ethnoscapes, (b) mediascapes, (c) technoscapes, (d) financescapes, and (e) ideoscapes." In fact, my own affectation (my own private Idaho one might say) is that the various sections here constitute imaginary physical elements of a virtual landscape. Thus one section in my mind is "the desert" while another is "the city of fractal towers" and so on. These remain unmarked (although—to summon the inevitable Derridean iterative— not unremarked).

Hypertextual Contours and Heterotopic Proprioception

AKiddleedivytoo [i.e., a kid'll eat ivy too].
—From a children's song

The imagistic approach begins with image but ends in translation. The metonymical approach would begin with translation—actually, the failure of translation—

and end in image, which is the only means of sustaining an ambiguous relation of polyvalent meanings.
—Donald Kunze, "The Thickness of the Past"

The feeling of being beyond oneself must of course reside within
one's self or else there would be no sense that one had projected
it beyond the self.

Everything depends upon that "it," although the allusive
structure of phrasing this sentence so in certain literary circles
marks it as a static modernism (viz. William Carlos Williams).

I have long ago accepted the fact that I am postpriorly a retro
modernist. (Viz Soja supra re con -text & -juncture.) And now I
have proof of this fact in print: "typical of both modernism and
[Joyce's hyperfiction] *afternoon* is the limited point of view . . . in
which the reader is denied access to any dominant hierarchic
structure, and therefore caught in a hetarchy" (Aarseth 1997, 89).

The first generation of hypertext fiction writers, pedagogues,
and literary theorists, as already noted above, were practiced in
systems where writers and readers of relatively bounded texts
("not infinite but very large" as Jay Bolter termed them) enjoyed
relatively rich interactive environments. With the emergence of
the image-driven web following the development of Mosaic, readers and writers take their place in a network of relatively
unbounded texts (where texts are understood in the pomo sense
that includes image—both dynamic and static, as well as sound
and collaborative interaction) that paradoxically only afford relatively sparse interactive environments. A third generation (Java-
pushed, one might say) is likely to trade off access to unbounded
space for enriched experience within commercially circumscribed
(intranetted or infotained) environments. These latter network
spaces might be not unlike glass-bottomed touring boats moving
soundlessly and invisibly above the sprawl and noise of the richly
populated but sparsely interactive net beneath, searching out the
neon tetra, Atlantis, and debris of fallen flights or provocateur
submarines.

The image works only if one imagines the glass-bottomed
boat outfitted with amenities: cocktail bar, avatar, shuffleboard,
search engine, captain's table, Real Audio.

The image can never work if it must be explained appositionally.

It is crucial to make the distinction between "work" and "doing work" as the third-generation, Berlin hypertheorist pointed out when the plump, cognitive scientist at the meeting in Hamburg (15:11:13 7Jan97) reported (regretfully, regretful he) upon the tenth anniversary of the publication of my hyperfiction *afternoon* that his data showed it did not work ("He proves by algebra," Buck Mulligan said, "that Hamlet's grandson is Shakespeare's grandfather and that he himself is the ghost of his own father") a fact that Espen Aarseth likewise confirms, that is, "*afternoon* does not contain a narrative of its own" (1997, 95) (I seem to be disappearing from Europe right beneath their eyes).

The image, in hypertext, always works appositionally in the biological sense of the growth of successive cellular layers.

"When I first spoke of living systems as autopoietic systems," Maturana says,

> I was speaking of molecular systems. Later, when I made a computer system to generate an autopoietic system . . . I realized it was necessary to make the molecularity of living systems explicit in order to avoid confusions. . . . a computer model takes place in a GRAPHIC SPACE generated by the computer, and this is why we did not claim to have a living system. . . . Yet it could have been proper to call all autopoietic systems regardless of the space in which they occur, living systems. (1991, 375–76)

In a usage of that term that most likely would set a geographer's teeth to grating, I have tried in a series of essays spanning nearly a decade to explore a notion of what I called hypertextual contour (I won't cite them all here given recent press reports that there are software agents that, speaking of contours, can generate a citational index, a sort of GIS—i.e., global information system—view of the intellectual landscape, which is sensitive to self-citing and flattens the cumuli of self-referential plateaus).

I meant nothing more (or less) than to describe the readers' (now the placement of this apostrophe, the surface structure, is significant) sense of changing change across the surface of a text.

I had in mind something less isobaric than erotic, the sense of a lover's caress in which the form expresses itself in successive-nesses without necessarily any fixation.

My most discrete formulation of this notion appears in the first chapter of this collection (though apparently, despite my fictional disappearances and hetarchical aspect, repeated here):

Contour, in my sense, is one expression of the perceptible form of a constantly changing text, made by any of its readers or writers at a given point in its reading or writing. Its constituent elements include the current state of the text at hand, the perceived intentions and interactions of previous writers and readers that led to the text at hand, and those interactions with the text that the current reader or writer sees as leading from it [and that are] most often . . . read in the visual form of the verbal, graphical or moving text. These visual forms may include the apparent content of the text at hand; its explicit and available design; or implicit and dynamic designs that the current reader or writer perceives either as patterns, juxtapositions, or recurrences within the text or as abstractions situated outside the text.

That formulation suffers beyond its seriosity (to use Woody Allen's term) from its fixity. In the first ellipsis in the block quote above I've excised a claim (here restored) that "Contours are represented by the current reader or writer as a narrative." The same claim is restored in action by Heather Malin:

when i'm writing a ht [hypertext] strange things happen. i am writing and creating, and i am fairly sure nothing coherent is going on. i am losing it. perspectives multiply as a realize the possibilities of my discourse. i see what is evolving, and i see contours, shapes, movements . . . i was not planning them, they were not of my authorial consciousness.

i try to follow the emerging currents; i start riding them and bringing them to some completion or exhaustion. but i am not sure that they are mine, although they are of my making. unless the text is writing itself, i am making it happen. somehow . . .

i end up stuck in the middle of my own movement. (1999, n.p.)

This stuck in the middle of movement is a giant step (albeit one suspended in stop-time like the children's game of "Mother, May I?") away from beginning-middle-end. In the space of what I've called the hypertextual "story that changes each time you read it," story becomes a matter of where you've been, de Certeau's already mentioned "act itself of passing by." In seeking to describe a notion of hypertextual contour I meant nothing more than what Malin suggests here: readers and writers do report the recognition of a form perceived outward from the middle of their own movement. This is the proprioceptive measure (proprioception is gut surveying, in which the surveyors levels spin around an inner sense of space and a sense of inner space alike) that literally embodies our on-board and in-born (and thus the first) global information system. A similar measure (in the musical sense of the Centrifug[u]e) seems to inform what John Pickles sees an explicit connection between ht and GIS.

> With the emergence of spatial digital data, computer graphic representation, and virtual reality . . . [t]he principle of intertextuality common to both hypertext and GIS directs our attention to the multiple fragments, multiple views, and layers that are assembled under new laws of ordering and reordering made possible by the microprocessor. (1995, 9)

Which is to say (again) that the gesture of the parenthetical, the dialectic, the thematic, the rhythmic, the fugal, the isobaric, the metonymic, the list, the link, the litany, as well as any and all other—whether em-dashed or no—appositional stitchery constitutes the space of hypertextuality.

More importantly this is space lived in, in the double sense of "in the body" and "habitat," both body and space being the arena of habitual, which by dictionary definition means "established by long use." In hypertext the forms of stories are quite literally established by use, not unlike the way that Sauer, in a gentler but no less threatened era for the science of geography, characterized the geographer's art as "distinctly anthropocentric, in the sense of the value or use of the earth to man" and, not unlike notions of hypertext contour, again and again located the actual (areal) for the world in the act itself of passing by:

> We are interested in that part of the areal scene that concerns us as human beings because we are part of it, live with it, are limited by it, and modify it. Thus we select those qualities of landscape in particular that are or may be of use to us. . . . The physical qualities of landscape are those that have habitat value, present or potential. (1968, 393)

Edge Wise

Even the lemellae of the desert slide over each other, producing an inimitable sound.
—Deleuze and Guattari,
"Treatise on Nomadology"

The critique of the "instrumentality" model doesn't explicitly take into account the aporetic, inconsistent effects of desire. It would be interesting to pursue the model of the pervert's reading of the link, to determine how it might cast instrumentality in ways that exceed mere utility.
—Terry Harpold, "Author's Note"

A knife is the figure of the inside out. I don't like to think about them. A well-sharpened blade repeats, and in the repetition anticipates, the contour of the inside out.

Where has the knife come from? The gesture is that of the *policier*. Its mere existence makes a mystery of known space. The knife is likewise history. The spatialization of history (its cut) makes a place for the play of reading and writing (which is of course the/a play of/on words) and the cultural margins (whose knife? the murderer's or the emperor's? do they differ?) that contain and represent them.

If hypertext is constituted by boundfulness—space that ever makes itself, slice by slice, section by section, contour by contour, never getting anywhere—then, like the spaces of our perception and occupation newly opened by GIS, it always both contains and simultaneously escapes the new orders to which it is subject. Each new slice of Zeno's GIS offers real estate for imperial economies yet likewise also opens a space of contestation for the foreigner armed now with the blade of the heretofore unexpected extents he natively inhabits.

In a section called "the silenced spatiality of historicism" within a chapter theorizing from the perspective of Foucault's heterotopia (and to which I am obviously indebted) Edward Soja defines historicism as "an overdeveloped historical contextualization of social life and social theory that actively submerges and peripheralizes the geographical or spatial imagination" (1993, 140). Indeed in a dazzle of boundfulness and a fearless succession of (self- and othering) citational edges, Soja in his chapter "The Trialectics of Spatiality" in *Thirdspace,* cites himself citing, in *Postmodern Geographies,* the protohypertextual urtext of "[the Argentinian writer Jorge Luis] Borges brilliant evocation of the Aleph as the place 'where all places are'" (1996, 56) as a locale where "*[e]verything* comes together," including

> subjectivity and objectivity, the abstract and the concrete, the real and the imagined, the knowable and the unimaginable, the repetitive and the differential, structure and agency, mind and body, consciousness and the unconscious, the disciplined and the transdisciplinary, everyday life and unending history. (1996, 57)

Hypertextuality considered in its widest aspect (whether web or three-space) has already fallen prey to an historicism of Soja's sort, ranging from the overdeveloped contextualization of so-called cyberspace by webs of interlinked advertisements for other adverts and search engines to the submersion and peripheralization of locality and embodiment alike—which thus far constitutes virtual reality. Of the latter, it is enough to say for now that VR perhaps provides the most obvious instance of how cyberspace exhibits what David Harvey calls "speculative place construction."

"Profitable projects to absorb excess capital have been hard to find in these last two decades," says Harvey, "and a considerable proportion of the surplus has found its way into speculative place construction" (1993, 8).

VR aside (and where else, pray tell?) there is likewise a kind of Cliffs Notes cartography of certain human interface specialists that too "submerges and peripheralizes the geographical or spatial imagination." Consider for instance the successive taxonomic slices of Fabrice Florin's (1990, 27–49) obviously earnest, if ultimately muddled, metaphors for data types in an essay titled "Information Landscapes." Florin suggests five categories (one hesitates to say spaces or entities); the following is my annotation of Dieberger's summary of them:

Collections of data that are represented as (literal) fields in the landscape. That is, fields with older data vanish to the horizon like certain kinds of legal testimony and lovers' excuses.

Interactive documentaries are visualized as a kind of village in which, one supposes, people come and go talking of Michelangelo. These are differentiated from the following.

Annotated movies that are characterized by their linear structure and so are represented (Heraclitus not withstanding) by rivers and (Al Gore and Bill Gates standing by) highways and so on.

Networks of guides that are visualized as other persons in the landscape, each one one supposes garrulous as streams and various as trees and always ready to help a stranger, Stranger.

Hands-on activities that range from simple games to complex simulations and that perhaps we ought to imagine as playgrounds, amusement parks, or Legos, i.e., little recursive versions of the same five data types.

This schema is of course merely an exteriorized version of the briefly popular "social interface" wherein we move from a desktop metaphor to include the kitchen sink. Here both desk and sink are leashed and taken outside for a walk. Because most HCI (human computer interface) specialists are innocent of history and naive as dollhouses or miniature railroads, the problems that accompany slicing the world into landscapes and that are the grist of postcolonial *policiers* don't bear rehearsing to them. Yet "reading 'the iconography of landscape,'" as Peter Jackson notes (1989, 177; appropriating the quoted phrase from Cosgrove and Daniels 1988) means "arguing from a world of exterior surfaces and appearances to an inner world of meaning and experience." To be sure one might long for an information landscape even half as rich in meaning and experience as Breughel's or at least one that could represent as much as Breughel's poet, the old master Auden, did of suffering and "its human position: how it takes place / While someone else is eating or opening a window or just walking dully along" ("Musée des Beaux Arts"). All this is, however, cut out of the world when taken as interface. Disembodied interface puts the world on edge, contourless. It is necropsy.

Against it there stands a different kind of cutting, the biopsic boundfulness of what Soja vis-à-vis the Aleph calls "all-inclusive simultaneity" that "opens up endless worlds to explore and, at the same time, . . . invokes an immediate sense of impossibility, a despair that the sequentiality of language and writing, of the narrative form and history-telling, can never do more than scratch the surface of [its] extraordinary simultaneities" (1996, 57).

Florin's information landscape would seem at first heterogeneous enough to suit a contemporary view of both simultaneities and socially constructed space, given its mix of landform, social scape, and potential human interaction. However, it is not coincidental that Florin comes to HCI from the world of television. This peopled landscape is sliced by the relentless transits of the surveying beam that sections scape and self and site alike.

Because the web links edgewise, I have said in chapter 3, it suggests that every screen is linked to another; thus on the web the true hypertext is the severing of one screen from another. Exclusion and inclusion interact, the outside defines the center (in the body this is called proprioception, that is, how the body perceives depth by its own depth, surface by its own: is the printed whorl where the finger ends or the world begins?).

There is a story in every slice. This is the story of contour, at least if taken as something other than a metric.

Joseph Paul Jernigan of Waco, Texas, convicted murderer, http://www.nim.nih.gov, the male Visible Human, comprises 1,878 slices for a total raw data of fifteen gigabytes, including MRIs, CAT scans, and photographs. The anonymous female Visible Human comprises 5,189 cross-sections. Jernigan's last meal comprised two cheeseburgers, fries, and iced tea. Jernigan was missing one tooth, his appendix, and a testicle at the time of his death.

Because of the computational power and dataspace he requires, the Visible Human exists only in networked iterations and dynamic representations. Because he is dead, he cannot be said to exist. Because he exists.

Space in the Singular(ity)

I placed a jar in Tennessee
And round it was, upon a hill.
—Wallace Stevens, "Anecdote of the Jar"

I take SPACE to be the central fact to man born in America,
from Folsom cave to now. I spell it large because it
comes large here. Large and without mercy.
—Charles Olson, *Call Me Ishmael*

Here is the parable of singularity, as told by the Bodhisatva of Ice.

My wife's son-in-law tells a story that for reasons of parabolical compression I usually retell as if it happened to him. He is driving back to his home in winter (for purposes of the parable I make this a cabin in the Sierra foothills reachable only by four-wheel drive along a logging road that is often closed; outside the parable he lives in Michigan, his name is Joe Moon) carting a large bottle of drinking water, the kind of oversized thick plastic jug that you sit inside a watercooler. He has been doing errands all day, it is a long drive from the cabin to town. At the cabin he puts the water jug up on his shoulder and begins to cart it up the path to the cabin when, suddenly, the sloshing stops and he feels it turn to ice in an instant, a solid block on his shoulder.

This is the singularity, the point where a system shifts states in perturbation (this sentence once had the word *causes* in it, once its subject was its object).

When I tell the parable I often say there is no single point, no zero centigrade, where the water becomes ice. Instead imagine the slow jostle of the Subaru (let's say) along the mountain roads, the rhythmic lift and dip of the jug on my wife's son-in-law's shoulder, the sudden shift when sloshing sleet is solid.

Yet the same is true of the parable, whether told here or in whatever there (the Philosophy Building at the University of Hamburg, for instance: you could mark one such actual telling so with coordinates of a GPS or the actual date of one of its tellings—January 8, 1997—yet there is no zero centigrade that constitutes a telling, no measurable parable, no historical story).

Already, above (where is this "above" I cite so casually?), the story has threatened to grow interstitially beyond the controlled sense in which I have let its multiplicities and simultaneities express themselves (as if they could be stopped from doing so). Some of these expansions seem as much musical (temporal) as spatial, for instance I was tempted (in fact, actually entered and then cut and repasted here) to add the phrase upside down to the following phrase now no longer above, now no longer the phrase I was tempted to add to, but another phrase with its own life here: "carting a large bottle of drinking water, the kind of oversized thick plastic jug that you sit upside down inside a watercooler."

Yet even the temporality here is multiple. At very least it is doubled, with the first temporality residing within the time (the prosody) of the sentence in which "upside down inside" takes on a certain pleasant rhythm and a comic (if not cosmic) joy; and the second the temporality (largely spatial) of *in illo tempore,* the vouching for the perceptible truth in a narrative that induces an audience or a reader having experienced an upturned jug of a watercooler to endow this story with plausibility (in fact Joe Moon has such a water jug in Michigan, in fact that is his name, in fact that, rather than Tennessee—or for that matter the California of Sierra foothills—is the state where he lives).

Yet where is the space of this part of the parable? Does it reside in the space of an imaginary foothill? the single topological surface of the torus upon the hero's shoulder? the notion of singularity as expressed by an only half-comprehending fiction writer afraid of being found out by a real physicist? the watercooler a reader imagines? the proprioceptive memory of the heft of an upturned jar?

It doesn't matter of course (it isn't matter of course). The space resides in the web of its tellings. Sometimes the space is literal (an obvious pun compounds into littoral) as in the paragraphical void after the framing device ("above" *in illo tempore*) of the Bodhisatva of Ice that enables one to occupy several narrative perspectives (where one is two at least: writer and reader); or the packed syntactic syntagma of "wife's son-in-law" that, almost ideogrammatically, tells its own story of postmodern succession, a story however "untold" here and thus one that truly (insofar as it is a story not included in the otherwise seemingly plentitudinous stories that open out from the parable) marks the bounds of the story that is told here.

This meta-telling is its own sort of verification and placement in several senses. You can see the space where the sentence ("above") was not edited for the rhythm of the inside upside down, and where (also "above" though elsewhere) it was so emended. You can see the self-reflexiveness of the story begin (parenthetically) to unravel. (A list can be likewise parenthetical, witness the ministory of the extended phrase that marks my anxiety speaking of singularities among physicists, geography among geographers).

Some such unraveling marks the space of plausibilities.

For instance in coming to write this story I began to think that the urgency that is gained by endowing the oral telling of the parable of the trip to town with the Sierra locale is somewhat lost, if not threatened altogether, in print where *Sierra* reads as a metonym for spring or melted snow (albeit romanticized—since a backpacker knows *Giardia* defuses some figures for clear water). I wondered, briefly, whether I should account for this, or rather whether such an accounting would gain in parabolical plausibility what it loses in sluggish scientism. (Words have this same, non- or not merely syntactic, trade-off: does "parabolical" or "scientism," each authenticated OED usages, cause the reader to attend to the sentence or to fall from it? and what of extended parenthetical musings? or the tokenized metaphor of defusing as a description of the action of a parasite upon a less-tokenized but conventional metaphor of mountain streams?)

A narratologist in the Hamburg audience was quite angry: "Space! Space! All you talk about is space! What's happened to time? A story is temporal." (January 8, 1997 Temperature 4C The Alster frozen.)

The hypertextual story is the space opened by its telling. And the space that it opens is called . . .

". . . House Again, Home Again, Jiggitty . . ."

This is the lost and found knowledge, the assurance of touch, head to foot. This is buoyancy, hazard, and waywardness—what it is to be at home, unhoused, ongoing. . . . Deprived of the elemental world—and who isn't, with a globe divided, the whole planet sectioned, roofed, cut and pasted—even its waters—what can a body do, if it is a body, but acknowledge, salvage, the elements in its own boundaries. Draw them out. Wring them out. Host. House.

—Janet Kauffman, *The Body in Four Parts*

This is about hypertext. "About" as in around&about. Or roundabout. Home&Page. Homepage. The enclosures of form that form us. Jiggity. Our boundfulness. Jag.

All around us, if you haven't noticed, homepages are disappearing, torn down or overwhelmed by online shopping malls. All around us, if you haven't noticed, homepages are appearing, persistent as toadstools in the parking lots of shopping malls. All around us home is here and not here, we are bounded by it, bound for it. Home is the heterotopic place "outside of all places" in Foucault's telling, "even though it might be possible to indicate their location in reality." And the name of their location (in reality) is house.

heterotopic≠house

The space in which we do and do not appear is the house. Foucault suggests that a history of spaces is a history of powers, including "the little tactics of the habitat" (1980, 149). The heterotopic strategy of making oneself the center of a text about boundfulness is house building. "With the house that has been experienced by a poet," Bachelard says, "we come to a delicate point in anthropo-cosmology. The house then, really is an instrument of topo-analysis; it is even an efficacious instrument, for the very reason that it is hard to use" (1964, 47).

Bachelard situates the poet's house so that in some fashion it reverses the flow of Kunze's fourth dimension with its "dynamics of moving from images to solid realities, i.e., the world in which human action becomes actual, the real and sharable world." For Bachelard the geometrical rationality of housal space "ought to resist metaphors that welcome the human body and the human soul." Instead "independent of all rationality, the dream world beckons" (1964, 47).

The architecture of an argument (a body, a hypertext, a readiness for something to happen) is not the same as the house (the world, a reading, the lines of desire) that sustains it. Kunze writes:

> "Architecture," while it requires artifacts to sustain it, is more like a readiness for something to happen. It forms the lines of desire and the defenses against danger. It crystallizes between our hunger for and revulsion with the world. Architecture is not identical with the objects it needs to sustain it. But when does a stone stop being a rock and start being the key of an arch? (1995, n.p.)

One answer is when we need it to be so. "The body is our general medium for having a world," Merleau-Ponty argues (1962,

146). Of his three possible worlds—including the biological that the body "posits around us" in "actions necessary for the conservation of life"; and the cultural in which "meaning . . . cannot be achieved by the body's natural means" and so "build[s] itself an instrument"—it is the mysterious third that suggests the space of heterotopic hypertext. Merleau-Ponty has no convenient container like biology or culture at hand for this sense that he baptizes a "core of new significance." As the middle world in his original text, it is presented as a literal movement from the biological to the cultural world, manifested through "motor habits such as dancing."

And so we come full circle (it runs in the family of unrelated Joyces) to Foucault's mirror of "mixed joint experience" where "I see myself there where I am not" and behind me I see Samantha spinning while the young men soar link to link overhead. It is here halfway spinning halfway soaring that I set up my ritual shed (this is about hypertext) and call it

homme sweet ho

Chapter 9

Forms of Future

This essay is shrouded in images of and allusions to Berlin, not merely because I first gave it there but also because that city serves, I think, as a locale for legitimate wariness about magical transformations, in this case the transformation of the book. When I first spoke about the matters in this chapter during a talk in an autumn in Berlin I was asked to talk about the particular transformation of the book that I know best, the state of interactive fiction. To say interactive fiction is what I know best, of course, does not necessarily mean that I am he who knows best about it, nor does it mean to suggest that interactive fiction is as yet anything but a marginal activity taking place at the sheltered edge of a plain scoured by winds of transformation.

I am reminded that the margin, whether the edge of the campfire or the hedge that shielded forbidden Irish bards, has been more or less the storyteller's place from the first. My friend Charles Henry, a great librarian and a technological visionary, often recounts his vision of the earliest storytelling technology. The cave paintings, he reminds us, could only be seen in patches of light from the rudimentary torchlamps—no more than fire upon a flat stone—held by our European ancestors of millennia ago. Those, too, were stories disclosed by littles and surely interactively.

In that spirit I will confine myself to what I can see, the edges of things illuminated by a brief fire in my hand. In doing so I will console myself with an understanding that prophecy is cheap in this age of suppressed memory. Today's market analyst and the technological guru tell the future by economic quarters but count on having their prognostications forgotten by the time the stock market closes that day. For most technologists the measure of the future is a soundbite, an animated gif, or a mouse click. I have written elsewhere that in our technologies, our cultures, our entertainment's, and, increasingly, the way we constitute our communities and families we live in an anticipatory state of constant nextness.

In this constant blizzard of the next, we must nonetheless find our way through both our own private histories and the cumulative history of our cultures. Not a history in the old dangerously transcendent sense, but a history of our making and our remembering alike: a history nearer to that which in *The Special View of History the* poet Charles Olson defines as "the function of any one of us . . . not a force but . . . the how of human life" (1970, 17).

Even so many are attracted to this subject because it promises the excitement of speed, the quickness of the present moment, the dizziness (or the Disneyesque) of next, and so I hope I do not disappoint with my slowness. Artists tell the future in millennia, a glacial measure that even (or especially) at the beginning of a new one is already haunted by the past, both the past gone and the past yet to be. The future of fiction is its past, though that future, too, is a fiction.

The emergence of a truly electronic narrative art form awaits the pooling of a communal genius, a gathering of cultural impulses, of vernacular technologies, and most importantly of common yearnings that can find neither a better representation nor a more satisfactory confirmation than what electronic media offer.

It seems self-evident that multimedia of the sort we see now on the web or CD-ROM is not likely to find a general audience. There is astonishing creativity everywhere (and I will point to some specific locales in a survey of interactive fictions at the end of this chapter) but there has not as yet emerged any form that promises either widely popular or deeply artistic impact.

Nor is it likely that a haphazardly swirling chaff of Java tools and plug-ins will suddenly reach a point of spontaneous combustion and bring forth a new light. The current state of multimedia does not repeat the case of the motorcar where widespread parallel technological developments led to a sufficient shift in sensibilities to make the mass-distributed assembly line seem a technological event threshold. The form of multimedia itself has no obvious audience, nor any obvious longing that it seeks to fulfill.

To be sure there will be electronic television, perhaps even the much vaunted, ubiquitous push technology that, once breathlessly championed by pseudo-religious cargo cults, techno-onanist publications, and infotainment empires, within a year fell from favor among the disciples of nextness. This fall from grace seems unsurprising since push technology as originally described was merely radio for the eyes in which infobits would flutter across the field of vision like papers falling from a virtual tickertape parade.

There will likewise be an electronic marketplace (perhaps there already is) for it is only an extension of the shopping mall with its shelves full of branded trademarks, surrounded by the architectural goulash of the gated suburb, and the holy shrines of the ATM card. The electronic marketplace will in this way parallel the course of the videotape rental industry in which an island of catalogs floats upon a sea of porn.

There are three general views about the failure of a true electronic form to yet emerge. Before I discuss them I wish to note that I have been quite intentionally using the term *multimedia* for the electronic television and electronic marketplace in order to distinguish such multimedia not merely from hypermedia but also from an electronic form yet to emerge but which has occasionally shown itself in almost magical, if incremental, transformations in our consciousness and indeed our sense of the real.

For now, though I will return to it later as a figure of more fundamental morphogenetic change, perhaps the image of Christo and Jeanne-Claude's *Wrapped Reichstag* can stand as a figure for these veiled changes, the pre-emergent and imminent forms of future whose edges push against the shrouded cloak of time like a baby's elbows push against a mother's belly.

One view of why a true electronic form has yet to emerge holds that we are in an age similar to that of the silent film and that a rich and powerful art form will emerge synergistically as the result of multiple, individual explorations upon the part of cultural producers coupled with simultaneously rising audience sophistication and expectations.

Yet the form of multimedia does not lead naturally from the marriage of eye and memory that film promised. Contemporary life leaves little time for those domestic and public mysteries of life lived in common that feed drama. Nor does multimedia provide the shadowbox for the psychoanalytic model of detached personality as does television. Multimedia neither extends the page into some inevitable dream of Technicolor longing to which its surface previously aspired, nor does it endow the unruly moving image with the staid conventionality of the page.

The second view about the failure of a true electronic form to yet emerge holds that authorship will turn from the creation of distinctly marked, individual stories to the creation of potentiated storyworlds, maintained and extended communally or by software agents that poll communal tastes. In such worlds individual audience members assume identities, spawn transitory narratives, and populate communities according to the logic of the storyworld, the accidental encounters of their inhabitants, and the story generation algorithms of software agents alike. The dream of the software agent and the storyworld is the dream of Scheherazade's mother, a longed-for happily-ever-after that is both outside the womb and yet no longer in the world. That dream doomed Berlin once before, before this rebuilding, the dream of a history outside history, a history at history's end. I think we all must be wary that dreams without ends do not summon the Reich of Virtual Reality, do not awaken the avatar führer.

The third view is perhaps an extension of the second. It holds that language slides inevitably toward image. From Jaron Lanier's sixties hippie, utopian view of unmediated, grokking communication through Virtual Reality to the network executives (of either the broadcast or inter networks) who see the web as packaging for a particular kind of targeted entertainment, not unlike the wrapper on a frozen egg roll, a Victoria's Secret brassiere, or the picture-in-picture headshots of interchangeable experts who appear over the shoulder of interchangeable infotainment news show hosts.

Total belief in the unmediated image is the behavior of cults. The Heaven's Gate cult knew what it saw beyond comet Hale Bop. Total belief in the unmediated image is denial of the mortality of the body. Yet outside the occult we live in a patchwork of self and place, image and word, body and mind. "Suppose we thought of representation," the philosopher and literary critic W. J. T. Mitchell suggests in *Picture Theory,* "not as a homogeneous field or grid of relationships governed by a single principle, but as a multidimensional and heterogeneous terrain, a collage or patchwork quilt assembled over time out of fragments" (1994, 419).

We will come to see (we have come to see) that electronic texts expose the patchwork ("expose" perhaps in the way of a photograph) and recall the body. "Suppose further," Mitchell says, "that this quilt was torn, folded, wrinkled, covered with accidental stains, traces of the bodies it has enfolded. This model might help us understand a number of things about representation" (1994, 419). The image Mitchell summons here is clear, the stained quilt is the Shroud of Turin, the bride's gift from her grandmother, the wedding night sheet, the baby's blanket. The image is clear but it does not proclaim its self-sufficiency.

The new electronic literature will distinguish itself by its clarity. It will seem right. I say literature because any literacy, even a visual or transitory one, expresses itself in a literature. Nor do I mean the kind of clarity that media purveyors speak about in terms of better authoring tools or more intuitive interfaces. I mean a new human clarity.

In the important special issue of *Visible Language* regarding New Media Poetry and guest edited by Eduardo Kac (1996), the French electronic poet and theorist Philippe Bootz quotes Jean-Pierre Balpe's assertion that because computer authors "do not question at all the notion of literature [but] on the contrary claim they belong to it and feed on it, the fact that they bring us to reconsider its nature and consequently its evolution seems unquestionable" (1996, 127).

The strengths of interactive fiction as a literary form increasingly seem to reside, quite curiously for me, in its realism, how truly it lets us render the shifting consciousness and shimmering coherences and transitory closures of the day-to-day beauty of the world around us. Hyperfiction seems equal to the complexity and sweetness of living in a world populated by other, equally uncertain, human beings, their dreams, and their memories.

Hyperfiction isn't a matter of branches but rather of the different textures of experience into which language (and image) leads us. Hyperfiction is like sitting in a restaurant in the murmur of stories, some fully known, some only half-heard, among people with whom you share only the briefest span of life and the certainty of death.

To be sure interactive fictions are an intermediate step to something else, but what that something might be is a question fit for philosophy. All our steps are intermediate. This one seems to be veering toward television, God help us, perhaps even television imprinted on your eyeballs. I put my trust in words. Media seers may talk about how we won't need stories since we will have new, virtual worlds, but soon those new worlds, too, will have their own stories and we will long for new words to put them into.

Do not mistake me. I am not suggesting that hyperfiction enjoys an obvious audience that multimedia lacks. I am however suggesting that language—with its intrinsically multiple forms, with its age-old engagement of eye and ear and mind, with its ancient summoning of gesture, movement, rhythm, and repetition, with the consolation and refreshment it offers memory— offers us the clearest instance and the most obvious form for what will emerge as a truly electronic narrative art form.

The new electronic literature will seem self-evident, as if we have always seen it and, paradoxically, as if we have never seen it before.

Berlin at this moment seems the ideal figure of what earlier above I called the astonishing creativity of an emerging electronic literature, a Berlin in which the cranes crosshatch a sky whose color, rather than being William Gibson's color of television, is not yet known, a sky whose expanse promises a new clarity *(eine neue Klarheit)*.

Despite the earnest impulses of government bureaucracies and the imperial appetites of transnational conglomerate capital, all of the busyness of the Berlin skyline—while not purposeless— is nonetheless to no purpose. This is good. We need to move beyond purpose, to what the monk and poet Thomas Merton calls the "freedom which responsibilities and transient cares make us forget" (1956, 17). We need to be free of technology to be free in technology. Like the overarching apparatus of our technologies, the scaffolding that crisscrosses Berlin is bandaged air. Beneath it lies the promise of new clarity, indeed even the unthinkable possibility of a *Kristall Tag,* an inversion of history, in which our world reforms itself as a globe of glass in which the fractures of the darkest nights are never again forgotten but rather where these healed-over fractures form a prism for a new light to shine through in all its differences. Through such a new prism the wounds of a world torn apart would both flow like tears and crystallize like roses at intervals in the way that the hearts of martyrs do under glass reliquaries in a cathedral.

The new electronic literature will seem old, as old as any human story, in its newness as old as birth.

The new Berlin heals over itself and in the process becomes itself differentiated by its own perception of gathering forms. The way in which a thing is both still itself and yet no longer itself is what Sanford Kwinter identifies as the singularity of catastrophe theory, in which "a point suddenly fails to map onto itself" and a new thing is born (1992, 58). This is, of course, the genius of Christo and Jeanne-Claude's *Wrapped Reichstag,* in which the thing seen is not the thing wrapped and yet evokes and insists upon it, and meanwhile the thing unwrapped is no longer the thing that was wrapped and yet promises to be what it was then.

This healing-over traces a circle like that of the Zen-like paradox of Alan of Lille (Alanus de Insulis), the circle whose center is nowhere and whose circumference is everywhere. In writing some years ago about the emergence of a city of text I cited Wim Wenders angelic vision of the great Berlin film *Wings of Desire,* in which angels walk among the stacks of a library, listening to the musical language that forms the thoughts of individual readers. Into this scene, shuffling slowly up the stairs, comes an old man, whom the credits identify as Homer. "Tell me, muse, of the story-teller who was thrust to the end of the world, childlike ancient. . . . With time," he thinks, "my listeners became my readers. They no longer sit in a circle, instead they sit apart and no one knows anything about the other" (Wenders and Handke 1987).

The new electronic literature will restore the circle as it always was and, paradoxically, as it never was before.

I suggested earlier that we live in constant nextness. Thomas

Merton speaks of the nextness of "Computer Karma in American Civilization" in which

> What can be done has to be done. The burden of possibilities has to be fulfilled, possibilities which demand so imperatively to be fulfilled that everything else is sacrificed to their fulfillment. (1983, 25)

The new electronic literature will bear the burden of possibilities in the way the earth bears the air.

Steven Johnson, the editor of the webzine *Feed*, recalls the passage in Walter Benjamin's essay "The Work of Art in the Age of Mechanical Reproduction," where "Benjamin talks rhapsodically about the cultural effects of slow-motion film" as an instance of how difficult it is "to predict the broader sociological effects of new technologies" (1995, n.p.).

"I've always liked this passage," Johnson says, "because it seems so foreign to us now, reading Benjamin fifty years later. If you imagine all the extraordinary changes wrought by the rise of moving pictures, slow-motion seems more like a side-effect, a footnote or a curiosity-piece."

Confronted by Johnson's observation I wondered instead whether Benjamin was right and we have missed the point of the technology. Perhaps we are all watching too fast. In his book of interviews Wim Wenders quotes Cezanne: "Things are disappearing. If you want to see anything you have to hurry" (1991, 1). Yet in another place Wenders says, "Films are congruent time sequences, not congruent ideas . . . In every scene my biggest problem is how to end it and go on to the next one. Ideally I would show the time in between as well. But sometimes you have to leave it out, it simply takes too long" (1991, 5).

The current generations of Berliners are, of course, citizens of the time in between and as such bear the responsibility that so many of us do in the constant state of changing change that constitutes network culture. The truth is that in network culture we are all from the generation of the time in between, and we too bear the burden of its telling, a process that, despite our technologies, requires constant generation and generations alike. One day Potsdammer Platz will be however temporarily complete. One day the world will lack a memory of what happened here, it is a storyteller's task to remember in the midst of dizzying change.

The new electronic literature will show the time in between, which is nothing less than the space that links us through our differences.

And so, as I turn finally to the brief survey of interactive fiction that I promised earlier, I hope I may be forgiven if I turn a critical eye toward the paradoxical lack of any obvious sense of what links us in these fictions. It is this lack of the *betweenus,* to use the word that Hélène Cixous coined, more than any technical lack, that momentarily stops us short of a mass electronic medium or a lasting art form. Nor do I exempt myself from this criticism. Although my hyperfictions are sincere attempts to negotiate whatever clarity I could find in link and multiplicity of voices, I sincerely believe I have as yet found nothing truly self-evident to show my readers. No new clarity, no new city of text beneath the cranes and scaffolds, no promised land, not even a wire frame Frankenstein awaiting the flesh of textural space.

And so as I begin this survey I would like to recall the modest intention with which I began this chapter, that is to say, what edges I think I can glimpse of forms of future. In the chapter I am less concerned with the sea changes that the new media labs and industries have made their business but rather to the ripples upon a surface that distinguish art from business.

Despite this doubled claim for modesty (which, like double negatives, doesn't not say I make no claim to see) I want to say something briefly about my most recent work, the web fiction *Twelve Blue,* which is available on web at Eastgate and elsewhere (1996b). I have often been critical of the way the web impoverishes hypertext. The web is a pretty difficult space in which to create an expressive surface for text. It seems to me that the web is all edges and without much depth and for a writer that is trouble. You want to induce depth, to have the surface give way to reverie and a sense of a shared shaping of the experience of reading and writing. Instead everything turns to branches.

With this fiction I decided to stop whining and learn to love the web as best I could, to honor what it gives us at present and to try to make art within the restrictions of the medium. *Twelve Blue* explores the way our lives—like the web itself or a year, a day, a memory, or a river—form patterns of interlocking, multiple, and recurrent surfaces. I've tried to use frames and simple sinking hyperlinks to achieve a feeling of depth and successive interaction unlike most web fictions. The idea is to put implicit links within the text and thus in a sense outside the interface and so to have the fiction echo with possibilities and transform the day-to-day, page-to-page rhythm of the web into a new music of swirling waters and shades of blue. So while there is only one explicit text link in any screen (and that one disappears when it is followed) the whole of the text is not only surrounded by the visual threads of its various linked narratives but threaded through with shared visions, events, and situations for which the reader's sensibility supplies the links.

The title screen contains a drawing of twelve colored threads running horizontally across a field of blue. The drawing came first, its threads creating a kind of score in the sense of John Cage, a continuity of the various parallel narratives. When the threads veer nearer to each other—or in at least one instance cross—so do their narratives. The twelve lines became months but also characters or pairings of them as well (that is, sometimes a character has her own line and another line she shares with someone paired to her, although not necessarily within the narrative threads). The twelve threads do not start with January at the top but rather November, the year of my year. I then made eight different cuts across the y-axis, though in my mind they were more fabric strips or something like William Burroughs's compositional cuts. These cuts are mapped on the opening screen in what is called an ISMAP, creating a succession of thematic sections.

Within these eight longitudinal strips the various stories take place and intermingle. Obviously however since narrative goes forward horizontally and time here is represented vertically, there is something of a displacement in which events along a single thread in fact violate the larger time of the characters' sensibilities. Thus the drowning deaf boy of the story floats across various threads through different seasons until his body surfaces at the end. Beyond this I gave myself some other simple constraints, for instance the already mentioned one of only one text link per frame and another of having every screen contain the word *blue*.

Meanwhile at the time of the writing I have barely begun another web fiction that takes place on an island inhabited by several historical characters (St. Francis, William Wordsworth and his sister Dorothy, and the engraver and book illustrator Bernard Picart). It is a novel about the relationships between word and image and the slippages as each lapse into each other. Parts of it are in a local pidgin of the island, whose name we never quite get, although the locals call it Banyan (or Yamland in some parts). In the words of the fiction pidgin also enters through occasional typos, which themselves enter the pidgin, since typos are thought to be sacred in this place, that is, divine inspiration, the devolution of the word, *logo* into *imago,* or so I think at present.

Before this brief excursus into my own work I invoked Frankenstein partly by way of beginning my survey of interactive fiction with that emblematic hypertextual figure as it is rendered and regendered in Shelley Jackson's extraordinary disk-based hypertext novel *Patchwork Girl, or, A Modern Monster,* (1995), a work attributed to Mary/Shelley and Herself. As I noted in an earlier chapter here, this is a fiction of continuous dissection, in which both Mary Shelley's monster and Frank Baum's girl of Oz are successively cut and repatched in the way of Xeno's paradox. Jackson's is a getting nowhere that gets somewhere and where, says the fiction's authorial voice, "I align myself as I read with the flow of blood."

Dissection and Frankensteinian cyborgization also informs the very provocative collaborative web work of Noah Wardrip-Fruin and others, titled *Gray Matters* (1997), itself a brilliant unbinding of book and body and the link each represents between creation and reception.

On the web I am currently very much taken by the work of Tim McLaughlin, whose language constantly meditates in the presence of image and mediates the nature of image. McLaughlin's work with the architects Thomas Bessai, Maria Denegri, and Bruce Haden for the Canadian biennial pavilion, *Light Assemblage* (McLaughlin et al. 1997), is an extraordinary exploration of how word makes place and place enables language. His *25 Ways to*

Close a Photograph (1996) perhaps most nearly approaches the self-evident quality that I have demanded of electronic literature, exploiting rather than working within the constraints of the web.

Although not strictly a fiction, I am very fond of *Memory Arena* and *Who's Who in Central and East Europe, 1933,* both done by Arnold Dreyblatt in collaboration with the Kulturinformatik department of the University of Lüneburg (1997a, 1997b), to which the German hypertext writer Heiko Idensen first introduced me. As Jeffrey Wallen notes in his introduction, this work takes "the ordinary and the bureaucratic . . . to a further extreme through their own logic of fragmentation, listing, juxtaposition, and leveling," giving us "a haunting glimpse of an absence" (Wallen 1997, n.p.).

Mark Amerika's overly earnest but nonetheless likable *Grammatron* (1997) is weighed down by a quasi-theoretical agenda, a perplexing nostalgia for cyberpunk, and the already discussed impossibility of multimedia. A similarly likable brilliance suffuses the work of both Marjorie Luesebrink and Adrianne Wortzel with a serenity of surface if not yet a fully new clarity. Luesebrink's *Lacemaker* web fiction (written as M. D. Coverley 1997) inside the also very compelling Madame de Lafayette site of Christy Sheffield Sanford is a variation upon Cinderella. Wortzel's *Ah, Need* (1997) turns the inevitable probing of surface that multimedia elicits to something more of an experience of linguistic surface. I likewise am drawn to the integumentary web explorations of the Kunst Brothers, a collaboration between artist Alison Saar and digital artist Tom Leeser, who see themselves as "bring[ing] together the analog world of physical materials and the digital world of moving images and sound into a transformation of existing spaces through the art of installation" (Leeser and Saar 1997, n.p.).

Finally I am especially fond of *Flygirls* (1997), the web fiction of the "WebWench," Jane Loader of *Atomic Cafe* fame. Its dusty rose to khaki trim retro look, its elegiac quality, and most of all its rich expanse and compelling writing are smart in the double sense of intelligence and style. This site seems an actual aerodrome but with the narrative spine of the race stretching over the rose-lit space, the links like lavender vertebrae.

My own feeling, however, is that the most provocative works are taking place outside the web in what might be called natural electronic spaces, the vernacular technologies of game engines, MOOs, and most especially the kinetic texts of electronic poetry where language finally finds its natural element in motion, not in a window but as a window, not as a single surface but as the aural, visual, and proprioceptive experience of successive surfaces. I do not think I am wrong to include hypertext fiction among these natural electronic environments, despite the current feeling in media and publishing and among certain critics that their time came and passed. This is hardly the literature of the present and will likely not be the literature of the future, and yet I am convinced that the literature of the present cannot continue without it and the literature of the future will not only encompass it but in some sense depends upon it.

An extraordinarily exciting international collaboration involves the Dublin-based but Derry-born writer Terence Mac-Namee, the electronic artist and programmer Eoin O'Sullivan in Derry, an American hyperfiction writer Noah Pivnick, and his colleague and coproducer Rachel Buswell (info at http://www.ulst.ac.uk/hyperfiction/Welcome.html). This group is in the midst of creating a fiction in the form of the Derry city walls, utilizing the Quake game engine as a locale for what they call networked co-readings. This story, which the authors describe as hypertext in architectural space, includes progressively disclosed texts, ambient sounds, and multiply inhabited story spaces that subvert the mythic war engine of Quake toward a literally dynamic consideration of the possibility of reconciliation. The fictional space invites the reader to explore walls and the link they represent between insider and outsider, reader and writer. Their fiction thus takes its place rather than takes place within a naturalized electronic space, not unlike how Judy Malloy in the early stages of *Brown House Kitchen* (1993) would set up space inside a room at Lambda MOO, the online textual virtual reality community, and begin to tell her stories, ignoring the protests, until the story made its own space.

Of my experiences of what might be called actual virtual reality thus far, I remember only one with a visceral excitement and longing: the experience of moving in and out of planetary spaces of text within a 2-D rendering of 3-D typographic space that I experienced in the work of the late Muriel Cooper together with David Small and Suguru Ishizaki at MIT's Visible Language Workshop. "Imagine swooping into a typographic landscape: hovering above a headline, zooming toward a paragraph in the distance, spinning around and seeing it from behind, then diving deep into a map," Wendy Richmond described it perfectly in *Wired* (1994, 184). "A virtual reality that has type and cartography and numbers, rather than objects—it's like no landscape you've ever traveled before, yet you feel completely at home."

Making space through and in and of language distinguishes the kinetic poets featured in *Visible Language,* whose work seems to me very much in the spirit of Muriel Cooper and her group. This includes Eduardo Kac's holopoems <http://ekac.org/holopoetry.html>, John Cayley's cybertexts <http://www.demon.co.uk/eastfield/in/>, E. M de Melo e Castro's video-poemography, Philippe Bootz's work on a functional model of *texte-a-voir* (1996), and most importantly Jim Rosenberg's <http://www.well.com/user/jer/> extraordinary body of theory and poetry leading toward an "externalization of syntax analogous to the externalization of the nervous system manifested in computer networks" (1996b, 115).

This is a call for a language outside itself, a language that goes out into the world. In his chapter "Walking in the City" in *The Practice of Everyday Life,* Michel de Certeau's spies this externalization in the figure of the wanderer who looks beyond "the absence of what has passed by" to "the act itself of passing by" (1983, 97). The act of passing by is Olson's history as the "how of human life." It takes place and makes place alike in the city of text.

There is a city of text and it, too, mutates and thrives beneath an umbrella of construction cranes and a crenellated skin of scaffolding, beneath SGML, XML, VRML, and HTML, inside the plug-in, the data stream, the web crawler, the game engine, the Photoshop filter, and so on. As with Berlin what matters most is not what life goes on beneath but what life emerges and in what light we come to see each other in the act of passing by.

Chapter 10

Paris Again or Prague: Who Will Save Lit from Com?

This is the second of two meditations about new media written in connection with talks in Berlin and then Prague and which I have come to think of as inversions of Tocqueville's reflections, that is, moving beyond democracy in America and, for good or ill, toward technocracy in networked Europe. I would like to believe that my own meditations have found a balance somewhere between the skepticism of the Yankee engineer in *A Connecticut Yankee in King Arthur's Court* and the baby boomer cluelessness of *Lost in Space.* What contemporary Europe offers to the American in Paris or Prague, however briefly, is a locale to observe the shifting moment of peoples lingering at the threshold of the intercultural disco, momentarily just outside the reach of the inexorable techno beat of nextness. Successive cultural revolutions and their technologies have left their mark upon the palimpsest cityscapes of Europe in what Lucy Lippard calls *overlay,* thus making them congenial spaces, at least at this instant, for slowing our descent into instantaneity.

Lately when I give a talk somewhere I have, much as in the last chapter, taken to apologizing in advance for any disappointment that might come to those who are drawn to such an event hoping for speed and newness. Despite being a hyperfiction writer, a first something or other in a medium that knows no firsts only nexts, I am also what used to be called a man of letters. Today, of course, letters seem slow to us. At first I considered writing that our sense of the slowness of letters is a result of our being burdened by memories of the halting first days of alphabetic life as schoolchildren when we formed letters with pencils gripped in uncertain paws, pronounced them with uncertain lips, scanned them with puzzled and uncertain eyes, not yet reading but merely deciphering, fingering words as uncertainly as if they were a line of ants creeping across the surface of the page. Suddenly, however, I realized that this inventory of childhood memories marks me further still as a slow man from a lost republic. For it is increasingly likely that many readers of this essay first formed, heard, even spoke and shaped letters as my two-year-old granddaughter Grace Moon does on the computer in the company of Reader Rabbit, summoning them up from pools of electronic light as if tracing the stream with a finger (which Ted Nelson claims was his formative memory of the essential hypertextuality of life), holding them quivering there at mousepoint like strands of melting pearls.

I realized that, for those readers and the Graces to come whose literacy is formed in and of light, words also suffer due to their slow unfolding in comparison to the visual image. However slowly an image discloses itself over a slow network connection, it nonetheless seems to dawn into significance in an instant. However swiftly it morphs or shifts or melts seamlessly into a video sequence, the image nonetheless presses upon us a sense of the distinct significance and insistence of its successive momentarinesses. The image means now: now and now and now again and again, despite our inability to mouth it, despite our unarticulated understanding as it shifts from now to now in what I have called the blizzard (but could as well have called the symphony) of nextness. In the face of this constant shifting from now to now and next to next we not only do not articulate the shift of image to image, we cannot articulate it. Images take on meaning for us in the blur between successive focussings, much as a frightened child on a carousel, for instance, picks out a parent's face from the blur of harlequin and sequin, chase light and twisted mirror.

If I am a man from a lost republic, it is one whose message is that we have seen lasting things in the blurring of moment to moment. If I am a man from a lost republic, its capital may be Prague, arguably the first city freed by networked consciousness. Prague is as yet, if only in the smoke-wreathed icon of its poet and playwright president, Paris again, the new, perhaps the last, republic of words, a new perhaps the last gasp of lit before com. In summoning the slowness of words a few moments ago I wrote a litany of their uncertainties, yet uncertainty is precisely the power that words offer the future as we fashion it both online and in ourselves. Indeed we will need to find words of courage and vision to sustain that elemental separation, which in attempting to create a model for "virtual realism" Michael Heim describes as "an uneasy balance . . . [between] the idealist's enthusiasm for computerized life with the need to ground ourselves more deeply in the felt earth affirmed by the realist as our primary reality" (1998, n.p.).

Who will save lit from com? There is a hubris to the question, an implicit nod to the theory of the "Great Man." "Hypertext has not yet had its Cervantes," Robert Coover wrote early on, though less than a decade ago, in the *New York Times* (1993, 9). And if it does, I thought at the time, it will not be hypertext.

The Great Man is the killer app embodied, and the last great man of literature, the unanimous choice of the unholy muses, the nine great men and one woman in New York who named the Modern Library one hundred best books of the century, was one with whom I share a surname but only an indirect lineage, through words. Perhaps James Joyce will be the last great man.

One should hope so on many counts. The first is the feminist argument, which I share, that not just the term *great man* but the implicit power relationship in this phrase is suspect. The Great Man myth spawns a hierarchy, which besides being strongly gendered establishes an infrastructure for cultural, political, and economical relations. It is a relationship of dependency, wherein the existence of a great man both requires and assures the existence of others to certify greatness, to curry, compare, and barter it quite literally against others. Moreover the dependency seems to shift poles, the school of pilot fish around the shark no longer depends upon his existence but rather the existence of sharks is formed from the swarm, in the way interlaced pixels resolve into form on the computer screen.

The shift of poles is, however, only seeming, only a semblance, for the great man myth denies the elemental interdependency that in Joyce's case kept him (albeit by inclination) linked to patrons and protégés alike for support and sustenance. The great man myth masks the anguish of the lesser man who grew up into or was grasped by the myth of greatness. It elides common life into categories. Thus a second argument against the great man theory is that it masks a fundamental democracy of the arts.

Democracy is a word still fresh in Prague. Yet it is also a word we hear less and less of elsewhere in the age of multinational corporate networking, resurgent nationalisms, and reactionary fundamentalisms. Democracy seems quaint to us and, like letters, a little slow. A Dutch internet business seeks to speed up democracy in quaint wooden prose on a website with backgrounds of scanned handmade paper:

> The Renaissance period was the rebirth of culture and science and transformed Europe in 300 years time. The current developments in the field of electronic media with Internet as its provisional epitome will change the world again within the next 30 years. Distribution of goods and information will change drastically and the access to knowledge will democratize. (http://www.siteways.com/)

Three hundred years or even thirty at three hundred megahertz has been made to seem intolerable. We cannot wait for the future in a culture of next, cannot track the past except as scroll or backlink. Thus the typical democratic answer to "Who will save lit from com?"—which is "You, dear reader, you will . . . " — is likewise quaint and what's worse susceptible to market measures that suggest a three hundred megahertz ability to take the cultural pulse of the populace and feed it back instantaneously in the form of images of individualized desires. This hortatory impulse toward the power of the individual (which we in the United States know as the fable of "How Anyone Can Grow Up to be President," though no poet has thus far done so there) is of course a miniaturized version of the great man argument. A similar version of this fable motivates United States policy vis-à-vis the worldwide infotainment industry. In its most benevolent form this policy, or really lack of policy, intends to create a multicultural surrogate of the Great Man. Any country can grow up to be America as long as it eats its Disney and takes a daily dose of CNN and wears McDonald's arch supports on its feet.

There is a weary resistance among artists, and especially literary artists, to postmodernist formulations. Yet it seems important to reconsider the postmodernist notion of master narrative as it transforms itself into a monster of democracy, what in Steven Johnson's words "are not the work of an individual consciousness, but rather the sum total of thousands and thousands of individual decisions." Johnson pins his hopes on submerged democracy, a shift from the fable of the great man to the myth of distributed communal consciousness. Writing provocatively in *Feed* about the shift from search engines to surf engines, he is especially excited by the "intoxicating" and "strangely fitting" idea of distributed intelligence in the surf engine Alexa's "What Next?" button (now built in to Netscape) where as

> the software starts to perceive a connection between the two web sites, a connection that can be weakened or strengthened as more behavior is tracked . . . the associations are not the work of an individual consciousness, but rather the sum total of thousands and thousands of individual decisions, a map to the web culled together by following an unimaginable number of footprints. (1998, n.p.)

The problem here is in the deceptive phrase "unimaginable" footprints. Architects use the term *desire paths* to describe those walkways worn into lawns outside the platted quadrangles, malls, and plazas. What makes desire paths compelling is that we *can* imagine in their cumulative footprints the democracy of particular presence and ordinary passage. If we believe the map has made the path, we lose our sense of the ordinary choices and footfalls that link us into life.

Who will save lit from com? A question posed in such categories is already answered of course. To the extent that human activity already falls under three-letter domains like *lit* and *com*— or even the oldest of the three-letter kingdoms and the mother of all technologies, *a-r-t,* art—the system of categorization already prevails. Such a system will save lit from com because for its purposes they are mere flavors, channels, portals, niche markets or tradestyles. There is economic value in sustaining marketable differences. The great lie of the alien term *content provider* is that metacontent becomes content. What is provided are categorical experiences; the custom-tailored suit is ready to wear, pret-a-porter.

Who will save lit from com? There is an old joke about the Boy Scout who sees an elderly blind woman at an intersection and who takes her arm to help her cross the avenue. The woman struggles fearfully but the Scout grips her arm still more tightly, speaking over her fears, moving her firmly across the avenue through the rush of traffic. "There, you see, there was no need to fear," the Scout says. "But I didn't want to cross," the old woman says.

Finally perhaps lit doesn't want to be saved, perhaps it wishes instead to cast its fate with synergy, the current dot.com buzzword, in which the literary takes its place in a spectrum of desires, from running shoe to foodstuff. In network culture, the symbiosis of the outsider culture and the master culture is heightened as the mainstream flows into a capillary web of—neatly niched—individual streams. Even if outsider channels are not themselves silted over or diluted by the overflow of multiplex culture, they are undifferentiated and undifferentiable among the many channels spawned by the mainstream. This might argue for a true symbiosis, of the kind of Hakim Bey's *Temporary Autonomous Zone,* in which the literary exists in interstices, unable to be dislodged from the multiple streams around it.

But we have had this hope before, in the middle days of the internet, when many of us imagined that a kind of capitalized authorship (capitalized in both senses) might, happily, melt away in a networked, collaborative age, and become recognized as merely a social construction in which the author function is reciprocal, negotiated, and distributed. In today's economy of bottled synergy such social constructions have a market value, however, and so we seem to be getting our wishes in the worst way. Commercial models of collaboration gleefully grant that authorship is constructed, a matter of market presence and placement, and writing is merely part of the mix. The global "military infotainment complex," to use Stuart Moulthrop's decade-old term, moves electronic media including the web rapidly toward command and control structures, that is, high-production-value, largely static, wide-bandwidth, predominantly unidirectional, so-called multimedia.

Meanwhile, as many commentators have pointed out, the emerging idea marketplace (including lit but also academe and polity) veers toward becoming a version of current film production or television where the writer/programmer (since the writer will also have to write computational objects and agents) takes a place in a collaborative production team. In the model of market synergy not only does the writer take place in a production team, but the written work takes its place in a marketing spectrum of products that are less products than images of them. In any case it isn't clear that the glass slipper of synergy will last through the royal ball.

As a self-styled man of letters I want to offer in place of a hollow and nostalgic resistance an electric certainty, an alternating current, which suggests that not just lit but com as well depends upon our ability to interrupt the flow of nextness with a sustaining sense of the ordinary. No one knows yet how to make electronic culture a popular medium. In contemporary theories of chaos, complexity, and emergence, the term *perturbation* describes the stone that interrupts the flow and transforms form itself into new flows. The ordinary is what transforms us. Or rather we transform ourselves through our awareness of ordinary being that has always been the business of lit rather than com.

I do not mean ordinariness here in the sense of either a certain kind of nervous literary minimalism or a nostalgic bourgeois realism. People have a complex sense of their own lives, which isn't ordinarily accounted for in popular art—they are capable of very complex relations. New media have to be faulted, ironically, for the failure to express that complexity in a way that people recognize. Nor do I want to suggest that the ordinary is solely the preserve of words, or to argue its case naively against the transformative power of image. Words create images in ways that, despite our increasing inability to describe either their similarity or difference, we know well. It is this sense of the shifting differences of day-to-day seeing that Kate Hayles finds in Christy Sheffield Sanford's webwork *Solstice,* whose "shimmering graphics . . . reflect and intensify the flickering nature of screen images . . . [and play] with the idea that light represents that which escapes verbal articulation—and yet it is the verbal articulations surrounding the images that make this idea clear" (1997).

Whether in light or life the seen surroundings form the drama of the ordinary. Among the most interesting electronic narratives I have seen in recent years are the environmental "intelligent multimedia systems" of Monika Fleischmann. To call what are essentially interactive video installations narratives (or, for that matter, cinema) is perhaps polemical, especially since I have seen Fleischmann's work only in demo tapes. Yet to the extent that the demo, as Peter Lunenfeld suggests on the maillist <nettime>, "has become the defining moment of the artist's practice at the turn of the millennium," it is always an exposition in time of a series of events. That is, the demo is a narrative. Fleischmann's self-consciously literary titles for these works, that is, *Liquid Views: Virtual Mirror of Narcissus* and *Rigid Waves: Narcissus and Echo,* reinforce their narrative bent. More importantly, however, not only is our experience of the tapes a narrative experience but also those whose experience is presented on the taped demos themselves seem to take on parts in consciously formed narratives.

In *Rigid Waves* viewers approach a framed video screen in which their own images seem to lapse and lag into creatures with their own will. "The visitor meets himself as an image in the image," Fleischmann says.

> He himself is the interface he is acting with. The approaching visitor will notice that he is changing the picture. The scene becomes photorealistic. This getting closer to real things is inverted due to the too-small distance existing when the spectator is leaving. The image trembles and becomes unclear. As in dreams, the images begin to speak through phase shift and extension. Shadows are produced, and finally the gestures become distorted. Movements get confused. Sequences of events are no longer near in time. The pictures seem to be oddly assorted. (1997b, n.p.)

What is most striking to the viewer of the demo tape is the variety of emotions and events with which these viewers greet what Fleischmann calls the "traces of . . . interactive dialog [that] remain visible in the picture." A woman in dark clothing greets her past self as if a sister or lover, they move together in complicitous rhythm and with evident laughter. An impish, heavy-set man attempts to trick the self he was instants before by the swift vaudeville movements of the Marx Brothers' mirror routine, familiar to film buffs. A professorial type in tweed seems unamused by his own history and cracks it into shards. This metafiction that I relate here, the stories of the woman in dark clothes, the vaudeville imp, and the fragmented man in tweed are, of course, tales that the external viewer produces. In fact, as I first wrote these comments, not having viewed the demo tape in some time, I was aware I might have been confusing the actors and the scenes. It does not matter. The interactions, as I have argued previously about the work of Grahame Weinbren and Jeffrey Shaw, present themselves to participants and spectators alike in layers of narrative. These simultaneous streams of private and public meaning are not unlike our common experience of space in ordinary life, whether over a kitchen table or a table in a cafe near the Vitava.

As Lunenfeld suggests of the demo, this common experience of the ordinary

> contains a multitude of contradictions. It portends to be about technology but demands the presence of the body. It speaks the language of progress but brings about an odd return of the cult value of the art object. It is both sales pitch and magic show. It is, in the words of advertising, the way we live now. (1998, n.p.)

Rigid Waves confronts the interactive participant with the way she lives now in company of the image of who she was however briefly before (and therefore at least partially still is as she and we now see her). Now is before the screen and then is on it. The succession of nextnesses is interrupted for viewer and participant alike by how each act.

Another work of Fleischmann's, *Liquid Views,* however, takes this interruption still further, erasing the gap, inserting it and us again into the present moment, interrupting the present with the present, interrupting the insistence of next with a liquid now.

Unlike *Rigid Waves,* in *Liquid Views* there is no split between now and then, off the screen and on. The present image is the present life and the present story is their confluence. It is almost a cliché of the new media industry that the future of com will consist of so-called story worlds, where audiences will share some sort of construction kits that provide setting, interactive characters, and the like that will spawn what in *Hamlet on the Holodeck* Janet Murray has called "procedural fictions." Almost alone among interactive artists Fleischmann has succeeded in such an enterprise, however briefly wresting lit from com, by providing a construction kit in which the story world is the one the visitor and viewer bring to it and the procedural fiction redeems that term from computation. The fiction is how we proceed in the face of what we see.

Liquid Views consists of a monitor mounted flat into the surface of a pedestal like a well and a large screen behind it and where what the visitor sees as she looks down we see behind her. Fleischmann's description is as simple as the work itself is haunting:

> The visitor approaches and sees his image reflected in the water—embedded in a fluid sphere of digital imagery. He tries to intervene, to touch the water surface and generates new ripples. Increasing the water movement too much, overstepping his limits, the viewer distorts his telematic reflection. The more he intervenes, the more his liquid view dissolves. After a time, while not touched, the water movement becomes calm again and returns to a liquid mirror. (1997a, n.p.)

On the demo tape the first visitor is a she and the simplicity and grace of her gestures untie the knot of what feminist film critics have called the gaze. We are not watching an objectified image but it is watching us, or seems to be doing so, with love and gentleness and a kind of seduction that is less cinematic than oneiric, the gaze of dreams. We slowly realize she is seeing herself and thus including us within both her gaze and her self alike. If pornography, as Susan Sontag suggested long ago, is a death pursuit, an obsessional nextness, which seeks to fill every opening until there are no more, then Liquid Waves is antiporn, the antithesis of pornography's inexorable computation. Stillness (a word that in English contains a sense of persistence and continuity, the opposite of nextness) opens present spaces in us.

Later in the *Liquid Views* demo a group of merry pranksters, not unlike Puck and company in *A Midsummer Night's Dream*, appear. In public talks Fleischmann has identified them as Dutch filmmakers who had come to do a documentary about the work and instead (or at least also) appeared in one. These fellows are aware of the mortal comedy within any toolbox and how in this one time moves on through rubber rollers within the tape machine, spooling from edge to edge of the dark box wherein our light appears. They make a little comedy of their reflections, something with a rise and fall and a sense of how, to paraphrase Chaucer, we fall together by chance, pilgrims all.

Pilgrims all, the way we live now is ordinary fiction, demo life, a multitude of contradictions, a new realism, one that truly takes into account the contingency, multiplicity, unfinishedness, and transience of our lives as we experience them. In ordinary life words are another toolbox, through which our lives spool. In *Krapp's Last Tape* Beckett's character coos our passing and his own in an unwinding word:

> Box three, spool five. (*He bends over the machine, looks up. With relish.*) Spooool! (*Happy smile. He bends, loads spool on machine, rubs his hands.*) Ah! (*He peers at ledger, reads entry at foot of page.*) Mother at rest at last. (1960, 13)

The oneiric qualities of visual images (always important for the son of a photographer as I am) are more and more intense for all of us whose memories are formed and embodied by Kodacolor and Photoshop and whose dreams are influenced, even infused, by what we see on all sorts of screens. Yet our lives do not lodge in technology. Krapp's mothering spool of the reel-to-reel tape recorder has given way to cassette and CD and to Real Audio spooled across the network. We can seal our memories into photos or DVD but they will never stay still. Images, to the extent they are the present and not the past, show the way for the future of images. As such they function not unlike the opacity of words to interrupt the inexorable flow of the next.

Fleischmann's art is a moving image in the double sense the word allows. It seems both something emotionally new and something very much imbedded in both cinema and verbal narrative. The conjunction of consciousness and the oneiric does suggest something like the aesthetics of movies but, as many critics suggest, movie aesthetics owe as much to the novels of the long-gone great men, from Tolstoy to the James Joyce, as to anything essential about the image. This is something that Jean-Luc Godard, the most bookish and hypertextual and oneiric of filmmakers, always understood, and that Wim Wenders does also.

The linear, the temporal, the syntactic, and the semiotic were the warp upon which the last great man, James Joyce, wove what was surely the first hypertext in the sense of Ted Nelson's first usage of that term, a text more text than any other text. Even so I do think Joyce would have been drawn to hypermedia not just for its oneiric but also for its civic sense, the mix of sonic and musical qualities, its quotidian complication, and perhaps even its glorious visual and design qualities that, even to failing eyes, would have seemed to rhyme with the dazzle of city streets, whether Paris again or Prague.

What links the one-minute commercial docudrama, the music video, and the web fiction if not the play of place and occasion? Can we imagine a fiction that honors the day-to-day complexity and shifting dramas of ordinary lives? Can we imagine a fiction that promises to close the gap between the fragmentary experiments of language and narrative that have characterized so-called literary or experimental fiction and the distinctly segmented consciousness of a larger audience who, from moment to moment, settle upon meanings for their lives in the intervals between successive accounts of their own or others' lives in several media and in various places.

Earlier I dubbed this new narrative ordinary fiction, having in mind the Ecclesiastical usage, which referred to the part of the Mass that remains unchanged from day to day, against which the daily readings come and go, or the division of the Roman Breviary containing the unchangeable parts of the cleric's daily office other than the Psalms. Michel Serres inverts a commonplace understanding to suggest, "It is a river that flows and yet remains stable in the continual collapse of its banks and the irreversible erosion of the mountains around it. One always swims in the same river, one never sits down on the same bank" (1982, 74). Fiction flows true and constant when the banks of ordinary life are undercut by the current.

In our ordinary lives we seek transitory closure, the recurrent momentary emergences of settled form, which Mark Taylor in *Hiding* connects with hypertext, where "the story or stories fold more than unfold" and are "never integrated, layers of the text are connected by . . . point[s] of intersection . . . that return repeatedly through all levels of the work. . . . Every time something clicks, a chain of associations is activated that allows the reader/viewer to shift between and among discontinuous textual strata" (Taylor and Miles 1998, 26). The transitory closures of ordinary lives need not be true and in fact are often fiction, the successive stories that Cixous describes as the "staggering vision of the construction we are, the tiny and great lies, the small non-truths we must have incessantly woven to be able to prepare our brothers' dinner and cook for our children" (1993, 63). Such fiction endows the ordinary feelings and occasions of our lives with the meaning we think to experience in them. We are our own stories and our stories are the stories of others.

Emotions are easy to come by under the regimen of next that is the world web, however much its wideness chokes in the narrow face of commerce. Even so we should be wary of nostalgia, wary anywhere but more so there where nextness compresses the natural succession of quotidian emotions that are not the ground but rather the flow that forms first the thin swamp of nostalgia and then, given time and a rich enough ground of being, the complex network of salt marsh and deep channel of the truer emotion of mortality.

It is easy to form a nostalgia for the early days of the web, some few moments ago, when the force of its presence was a largesse, a haphazard weave of idiosyncratic "home" pages, naive graphics, and cobbled-together texts ranging from refrigerator maintenance manuals to dubious editions of Shakespeare to the most intimate journals, disclosing everything that the surface views of sites like Jennie-cam are, despite their dreary tabloid voyeurism, unable to show, no matter how many layers of pretty pink their putative subjects strip off or open up to the gynecological camera lucida.

It would be a mistake to think this five-year-ago past a paradise. Even at its inception there were complaints that the web was clogged with self-conscious maunderings and self-indulgent claptrap. The cry then (O can you recall those distant days of months ago!) was for quality, which somehow meant the reemergence of canonical arbiters, suddenly however supposed to be not only multiculturally benign and gender unbiased, but also not merely tolerant of but champions of new genres and forms, ready to see their virtues, unfazed by squinting into the light of the screen after the tallow light of the study. There was a hope that refereed sites, filters, and software agents, not to mention new industries, would borrow upon the spirit of the web's formative donnée and particular presence to create new forms and forums alike (the word *content* perhaps floated by then but only as a dry leaf among sweating blossoms).

We know what happened next. And next. And next. And next. That is, in fact, all we know. Next after interminable next. Between the oppressively expansive naïveté of the web's first offerings and the commercial glaze of its present state, something has not so much been lost as colonized, shunted off to the nascent villages at the outskirts of geocities or machine-stitched into the metallic threaded and mirror-paneled patchwork that drape online portals.

Against this occasionally comes something still uniquely, one is tempted to say heroically, quotidian as "A Journal for My Child" by Deirdre Grimes (1999). I offer that site here not as any evidence of a breakthrough text, since breakthrough only gives credence to the culture of nextness, but rather as work fairly considered, beautifully rendered, a gift in space and of time and the body, what the web was once and could be, perhaps even is, still flowing.

On this site, at this sight, faint violet (too busy, too Java-burdened) frame windows, a grid of weeks of pregnancy from five to forty-one on the left (with labor and motherhood appended underneath) becomes a quilt, scanned handmade paper notes and illustrations ("Motherhood—Week 1—April 3rd to 9th," for instance, begins with a tender drawing of a nursing woman's breast, a drawing so sweetly rendered that you wish that all the boys seeking breasts on the web might see it and know that a line is love and that "why, how, who, what" we write is, in Hèléne Cixous words, "milk. Strong nourishment. The gift without return" [1991, 49]).

This gift begins at one of those points of no return and with its last line a typographic error of the sort that the poet Charles Olson treated as direct communications from the muse and so always let stand:

> In July of last
> year I
> discovered that
> I was pregnant.
> I was
> unmarried,
> insecure about
> my partner's
> feelings
> towards me,
> immature as a
> woman and I
> had a year of
> collage left.

The collage/college here is liminal, a "context" in the root sense, a conjoint text "woven with" one's life and thus, ultimately, others lives:

> THIS JOURNAL
> IS
> COMPLETELY
> UNEDITED
> AND
> UNCENSORED.
> IF YOU KNOW

ME "IN REAL
LIFE" AND
FIND
YOURSELF
HERE, PLEASE
RESPECT THE
FACT THAT
THIS WAS
ORIGINALLY A
PRIVATE
JOURNAL AND
TRY NOT TO
TAKE
ANYTHING
OUT OF
CONTEXT.

What can be faulted is not that her telling invades others'
lives but their failing to see their lives in the context of her telling.
Grimes's genius is the leaving in of context, whether in a view of
a Dublin park with

> people baring whatever inches they dare bare and soaking up
> the heat with all the pleasure of cats. Foreign students with
> backpacks sitting in large organised looking circles on the grass
> and a very faint smell of burning like everyone is getting slowly
> cooked. (July 31)

or in the larger context where her web reclaims and inscribes the
Muses' true space, the "significance of the moon" that is the
embodied word and image:

> Well I wrote into the made journal for about 2 weeks then the
> paper ran out, since it takes me a long time to find paper that
> I'm happy with I thought I should keep recording the days any
> where, this is the easiest. The strangest thing about all this is the
> difference between this and my last journal, up to this my vulva
> was the source of two things, pleasure and blood, in 6 and a half
> months time a baby is going to come out of it, will I ever look
> at it the same again? It also bothers me a little that I won't have
> a period or the significance of the moon phases for 9 months. I

feel abandoned even though I know that even that 9 months is dictated by the moon. (August 30)

Ultimately this site has no ultimate form (nor should we expect one of a genuine journal of a new life) and thank God no "content" in the sense that divorces that word from any body, although there are sequences of traditional, temporal drama, as these two textually separated entries from the "Labour" section make poignantly and powerfully clear:

> As the midday news began, the anesthetist arrived to administer the epidural. Eugene was told to leave "get a cup of coffee," I envied his place in the ordinary world, buying an ordinary cup of coffee. While they gave me the injection, which is usually given before the strong contractions, I had to stay perfectly still and could not use the gas. I thought I was going mad, I remember looking out the window at the rain and some part of me was out there, flying over limerick, feeling the rain on my face.

> I realise now that this is where the pregnant woman I was died. There were two new people in that room a new mother and a new son. Strangely I don't think Eugene noticed the difference, he kissed the new mother like he had always known her and she felt wicked for taking the dead woman's lover.

The woman who died gives birth to the new woman who writes where it is easiest, where one needs no paper, only light and power. Hers is the story of flying, feeling the rain on one's face. There is no ending, although as of this writing her last entry suggests an anxiety about having one thrust upon her:

> Last night I dreamt that my head tutor put me down on the sign in sheet as "finished college" and wrote "possible fail" after this, how's that for anxiety? ("Motherhood—Week 4—April 24th to 30th")

I want to avoid the temptation of the rhetorical turn one takes at such times ("for this viewer, at least, this is no failure"). It seems better to acknowledge the truth that all we have before us are the possibilities. To be something other than mere nextnesses these must also include the possibility of failure as well as the words and visions with which we mark possibility and weave it into the gifts of bodily life and the ordinary rhythms of the moon's significances.

Beneath the moon every city, whether Paris again or Prague, is always new. The Paris I mean to evoke here is the Paris of transformation whose presiding genii, according to Alice B. Toklas (at least as she is represented by the first of them), were Gertrude Stein, Pablo Picasso, and Alfred North Whitehead. Stein is the figure of the spatial word, Picasso of the temporal image, and Whitehead of the actual occasion. We can think of these three as patron saints of a new media. Together this perfectly imperfect trinity are totemic of the transformed ordinary that I am characterizing as Paris again or Prague. In the work of each of them the quotidian, in words images or events, slows and forms the flow of nextness into ordinary significance.

The first patron saint, the doyenne of syntax and recurrence, Stein, in her life and writing alike, prefigured Mark Taylor's sense of the hypertextual and thus the network as "chain of association . . . between and among discontinuous textual strata." Textual strata in their opacity and recurrent excess give way to images. The web artist Christy Sheffield Sanford rightly I think links her own sense of the computer page as "scenic . . . less akin to the still life or portrait and more akin to the landscape" to Stein's discussion in *Lectures in America* of the play as landscape. "When one sees a landscape," Sanford paraphrases, "one takes it in all at once" (1997a).

Of Picasso there is already (and probably was from the first) too much said, but that is precisely the point that makes him a fitting prototype for the complex of meanings surrounding overly saturated images, especially in electronic culture. His forms slow down the emotional simultaneity of landscape and instead become inscapes or spaces of potential inscription. From the first the circus of publicity descended upon Picasso's inscapes. We never see them or him and can only mark them in aftermath, after the tents are down and the caravans move away, reading rhythms of significances in the trampled rings of what Kate

Hayles calls "the verbal articulations surrounding the images." The same is true with the constantly passing events of our ordinary lives. The excess of Picasso, which is to say the networked image, both tests and hones our belief that we can say something. Belief is something we must now more than ever be parsimonious about. We can't give over to belief too easily. Too much vies to fill the slipping gap between once and will, past and future, in ordinary life. To redeem the slowness of event requires disbelief, a philosopher's task.

Of this totemic trio Whitehead is the least known, the philosopher of prehension and concrescence, strange words even to a native English speaker, and yet again the right words for a network culture. Prehension, with its echo of the word prehensile, the primitive grasp of the ape ancestor, is an unconscious awareness of form, a disposition, the kind of "feeling" that isn't the grasp itself but grasping at. Prehension is the kind of feeling one means in the ordinary usage of "I have a feeling this will turn out well (or badly)." Concrescence is its result though not its effect in any causal sense, not the outcome nor any thing exactly as much as process given over to entity.

This may seem metaphysical (more properly epistemological) gobbledygook, yet it provides a language to describe our experience of how successive present-tense events, for instance a series of hyperlinked web pages, weave themselves into transitory closure, a recognizable and ordinary, if not definable, coherence. As something both more and less than understanding, something that cannot be sold to us, transitory closure is what a decade ago I characterized in defining constructive hypertexts as a "structure for what does not yet exist."

Our lives are such structures. Convinced that our hope is in the ordinary, our senses and our words combine to form and commemorate what we imagine to be our lives. Even so we can't give over to belief too easily. Even so we are never consoled. Yet what shall we do, we wonder, what is to be done?

At the beginning of a previous century, in Paris again, in word and image, a number of figures faced such questions. At that time, certain structures for what did not until then exist emerged in both word and image and have served us for nearly a hundred years. Miss Stein has Miss Toklas describe these changes in a different totemic trinity (yet in which two figures unsurprisingly persist, in this telling the philosopher however giving away to another painter):

> It had been a fruitful winter. In the long struggle with the portrait of Gertrude Stein, Picasso passed from the Harlequin . . . period . . . to the intensive struggle which was to end in cubism. Gertrude Stein had written the story of Melanchtha . . . the first definite step away from the nineteenth century and into the twentieth century in literature. Matisse had painted the Bonheur de Vivre and had created a new school of culture which was soon to leave its mark on everything. And everybody went away. (1933, 54)

We do, you know, we go away. That's what we do best. That is what we can do. We go home to our lives, not as great men and women but as ordinary people, setting ourselves as stones among the flow of nextness, turning ourselves and thus the world to liquid views and the moon's significances.

Chapter 11

My Father, the Father of Hypertext, and the Steno Whose Impulses Were Interrupted by a Machine: Vannevar Bush and the War against Memory, a Theoretical Heretical Narrative

Part 1 or One Part: *La lutte continue*

This has not been a scientist's war; it has been a war in which all have had a part. The scientists, burying their old professional competition in the demand of a common cause, have shared greatly and learned much. It has been exhilarating to work in effective partnership. Now, for many, this appears to be approaching an end. What are the scientists to do next?[1]

For anyone who has written, and written about, hypertext for the last decade or more it is hard to avoid approaching Vannevar Bush's essay "As We May Think" with something of the same question in mind. What can we say next? The orthodox accounts (Bush to Engelbart to Nelson to everything else), my own included, take on the old testamentary feel of the Book of Numbers: "Of the children of Manasseh by their generations, after their families, by the house of their fathers, according to the number of the names, from twenty years old and upward, all that were able to go forth to war" (1:37). The Memex was the first skirmish in the war against memory. War is over if you want it, John and Yoko sang. When Johnnie comes marching home again, taroo,

---

1. The extended citations in the first part of this chapter (as all otherwise unidentified citations in the second part) are from Bush 1945 and are reproduced there with the permission of the *Atlantic Monthly*.

taroo. Are you looking for connections? Then look around the edges. Soldier home is an old story in this country, although in our time Hemingway perhaps told it best: no one remembers, after all is said and done, the enemy is the next itself. It's the one thing we can kill off with constant success: next, next, next, next . . .

But every time you kill it off, there it appears again, that is, next. Next was once. The so-called posthumous festschrift for Bush held at MIT on Thursday, October 12, and Friday, October 13, 1995, bore a full subtitle worthy of the five-year plans of the Reds of my postwar youth: "A Celebration of Vannevar Bush's 1945 Vision, an Examination of What Has Been Accomplished, and What Remains to Be Done." These remains are what links us. If every link is a next, every memory is a was. The question becomes which side are you on? Was or next?

> It is the physicists who have been thrown most violently off stride, who have left academic pursuits for the making of strange destructive gadgets, who have had to devise new methods for their unanticipated assignments.

The sweetness of it strikes you only after the incomprehensible understatement. "Strange destructive gadgets" left imprints of the flash-fried flesh of real humans in the form of shadows on the cement of Nagasaki. Gadget is the language of grandpas, and this phrase strikes with the slow force of what we imagine were more innocent times (I was born in the November after Armistice 1945, a month before this essay). It's as if we witness a Norman Rockwell drawing of a kindly old doctor holding a smoking machine gun. Go! what a gadget! what "unanticipated assignments" indeed! Makes a man long for a laser-fired pipe, an atomic hearth, an electric study.

There is a perplexing nostalgia for the future throughout the essay, a nostalgia done up in grandfatherly evocations of the dangdest gadgets, such as the "lump a little larger than a walnut," the Cyclops camera, which "the camera hound of the future wears on his forehead." But don't dare doze off with visions in or on your sugared head: Grandpa's got the numbers down: "It takes pictures 3 millimeters square, later to be projected or enlarged, which after all involves only a factor of 10 beyond present practice."

You knew that, didn't you, sonny?

Something of the same tone of gadgetry strikes you, a mere two years after, when you read on the web page for the 1995 MIT fifty-year celebration of this essay that "The symposium will be broadcast via the Internet Multicast system MBone."

MBone MBone where you been? Round the block and back again.

Next was once. Lost gadgets of yesterday like MBone seem to float like walnuts along the streaming video that leads us to tomorrow.

> Science has provided the swiftest communication between individuals; it has provided a record of ideas and has enabled man to manipulate and to make extracts from that record so that knowledge evolves and endures throughout the life of a race rather than that of an individual.

The initial claim simply isn't true, unless by Science one means to subsume its subjects, otherwise known as the living world. As we begin, so we may think, to decrypt (the verb is laden and hides a grave) the genome, we begin to see that the swiftest communication between individuals is the moment of conception. Moreover that transmission is high bandwidth and multicast, often containing messages from the whole race of races of men and women, that is, "knowledge [that] evolves and endures throughout the life of a race rather than that of an individual."

The glance of a mother (or a lover) is often as fast as the genome or at least the fiber optic link. So is the dance, the drum, the ideographic trace of the calligrapher's brush, the film frame, the slash of paint, the almost inaudible hush of the caesura at the center of a poem's metrics. These, too, technologies in the radical sense of that word: *tekne,* the tool, the touch.

While it doesn't do, at the start at least, to fault Bush for his penitential fervor and positivist glee as he looks up from the salted wasteland and imagines fruit trees, it is interesting to note that, if this essay marks some margin from which we measure the age of information, from the beginning there was the reluctance (or perhaps the simple inability), which also endures these fifty-plus years afterwards, to acknowledge the swiftness and transcendence alike of art as a mark both of individual and the human race. The first "cheap complex devices of great reliability" was the mark, whether cave painting or cuneiform, and ever since and by its nature "something is bound to come of" the mark, even inscribed in light.

> A record, if it is to be useful to science, must be continuously extended, it must be stored, and above all it must be consulted. Today we make the record conventionally by writing and photography, followed by printing; but we also record on film, on wax disks, and on magnetic wires. Even if utterly new recording procedures do not appear, these present ones are certainly in the process of modification and extension.

I remember Science. It showed up sometime in the 1950s and lingered, really, into the late seventies before it disappeared in favor of Information. Science seemed made of tile and glass and was sheathed in white cotton, smelling of alcohol. For awhile in the fifties and sixties it floated around the world in the company of suborbital monkeys and Russians in tight-fitting aviator leather and then (finally!) boys from Ohio, done up in science-white suits, unable quite to recite the poem they tried to memorize but nonetheless tramping firmly on the moon in silver boots. Information seems made of gas, constantly expanding, seemingly uncontainable, like the helium of the *Hindenberg* or the lost balloons of children at zoos. Information shares Science's urgent need for continuous extension and storage of the record but it isn't clear at all whether the record must really any longer be consulted. At the ACM Hypertext '96 panel "Future (Hyper)Spaces" in Washington, D.C., the hypertext researcher Andreas Dieberger said to great applause and a murmur of general recognition among cyber-boys, "I don't read anything on the web anymore. I just check out the links and mark the ones I want to come back to later. Though I never do really."

Information floats away like balloons. The web is like watching someone talk underwater, the bubbles floating away like the chuckles of characters in comics. Science cruises through the scene in pristine white exploratory submarines *(mais Jacques Cousteau est mort)*.

Will there be dry photography? It is already here in two forms. When Brady made his Civil War pictures, the plate had to be wet at the time of exposure. Now it has to be wet during development instead. In the future perhaps it need not be wetted at all. There have long been films impregnated with diazo dyes which form a picture without development, so that it is already there as soon as the camera has been operated. An exposure to ammonia gas destroys the unexposed dye, and the picture can then be taken out into the light and examined. The process is now slow, but someone may speed it up, and it has no grain difficulties such as now keep photographic researchers busy. Often it would be advantageous to be able to snap the camera and to look at the picture immediately.

I remember the brown gum of the early Polaroids (I should use a trademark here) and how slow the instant was, not unlike what I back in chapter 3 called the ostrich eggs of someone's slowly hatching gifs on their homepage. Wim Wenders quotes Cezanne: "Things are disappearing. If you want to see anything you have to hurry." There are digital cameras now in which you can "quote" earlier photos as easily as someone scoops up an already developed sentence in a word processor. My own father was a professional photographer during strikes at the steel plant. He would have liked the whole idea of the string that takes the pictures running from the walnut and down the sleeve and he surely would have known about "diazo dyes which form a picture without development." Probably he would have seen something about Dr. Bush in *Popular Science*. I'm not sure how much he knew about Bush from wartime reports; 4F himself because of a bad back, he was more interested in news of his brothers, one slogging through the South Pacific, one at the Battle of the Bulge.

My father wouldn't have been brave enough to take up Bush's

rhetorical dare that "it would be a brave man who could predict that" the process of rendering instant photos "will always remain clumsy, slow, and faulty in detail." No one would.

And he wouldn't have been canny enough to see the possibilities of micropayments once "the microfilm Britannica . . . cost a nickel, and . . . could be mailed anywhere for a cent." Ted Nelson did. The transclusive hypertext he invented was a complicated cash register and royalty scheme, one that the Seattle lawyer's son is still dying to build for real.

Bill the Gateskeeper owns all the images that he can find, brown gum, diazo dye, dry photography as fast as television.

My father wouldn't have liked him. It's all too abstract really, too much like air. Information. My father was a guy of Bush's time, a gadgeteer, taken with Science. He would have liked the long discussion of electronic beams in vacuum chambers that seems beside the point to us now.

But after all isn't hypertext, by definition, beside the point? "That," as Bush said, "introduces the next aspect of the subject."

> At a recent World Fair a machine called a Voder was shown. A girl stroked its keys and it emitted recognizable speech. No human vocal cords entered in the procedure at any point; the keys simply combined some electrically produced vibrations and passed these on to a loud-speaker. In the Bell Laboratories there is the converse of this machine, called a Vocoder. The loudspeaker is replaced by a microphone, which picks up sound. Speak to it, and the corresponding keys move. This may be one element of the postulated system.
>
> The other element is found in the stenotype, that somewhat disconcerting device encountered usually at public meetings. A girl strokes its keys languidly and looks about the room and sometimes at the speaker with a disquieting gaze. From it emerges a typed strip which records in a phonetically simplified language a record of what the speaker is supposed to have said. Later this strip is retyped into ordinary language, for in its nascent form it is intelligible only to the initiated. Combine these two elements, let the Vocoder run the stenotype, and the result is a machine which types when talked to.

Voder, Vader, Father . . .

It's hard not to laugh at the story of the Voder and Vocoder. You start to wonder whether the two Vs had met, and if they perhaps knew the steno who stares languidly around the room with the disquieting gaze and disconcerting device. Nerdy guys always feel strange under the gaze of dreaming women. Poor Bush seemed daunted by the steno's dreams and her ability to summon in "a phonetically simplified language a record of what the speaker is supposed to have said." What do women want? Freud supposedly asked. Darth Vader knew, he saw the dry photographs of Princess Leia printed holographically in the air. He didn't like what he saw.

We've seen the dry photographs of Princess Di printed in the air. Now we claim we never liked what we saw.

Why does this whole section of Bush's text seem a comic love story of man and machine? "Our present languages are not especially adapted to this sort of mechanization, it is true," Bush says. He is no longer present.

Besides coining *hypertext,* Ted Nelson also coined the term *teledildonics,* for computer-assisted distant loving. "For mature thought," says Bush, "there is no mechanical substitute. But creative thought and essentially repetitive thought are very different things. For the latter there are, and may be, powerful mechanical aids."

This is a story of Jack and Diane, said John Cougar Mellencamp, who began the process of progressive taking on and off of names that Prince only later turned into an art. What does all of this have to do with anything? you may ask.

I honestly believe that hypertextuality is about sexuality. As Bush says:

"Such machines will have enormous appetites. One of them will take instructions and data from a roomful of girls armed with simple keyboard punches, and will deliver sheets of computed results every few minutes. There will always be plenty of things to compute in the detailed affairs of millions of people doing complicated things."

He most likely imagined the complicated people as men.

The needs of business, and the extensive market obviously waiting, assured the advent of mass-produced arithmetical machines just as soon as production methods were sufficiently advanced.

With machines for advanced analysis no such situation existed; for there was and is no extensive market; the users of advanced methods of manipulating data are a very small part of the population. There are, however, machines for solving differential equations—and functional and integral equations, for that matter. There are many special machines, such as the harmonic synthesizer which predicts the tides. There will be many more, appearing certainly first in the hands of the scientist and in small numbers.

Every now and again—among the reports of false computer viruses like Good Times and the electronic versions of the kinds of Xeroxed jokes that used to be taped up in office mailrooms and above watercoolers before the days of Starbucks—email brings someone's belated discovery of the same hoary, stock file, an anthology of bad technological predictions, including the reputed prediction of an IBM exec that only a hundred people or so (I never really read the email with these quotes) would ever want computers. Bush's modest little claim here that "the users of advanced methods of manipulating data are a very small part of the population" probably should be on such a list.

Almost everyone knows how to do a Boolean search for "myfirstname mylastname porn sex baseball free etcetera." Who has not experienced the joy of searching thirty million web pages in a nanosecond (or the nearest we humans are allowed to experience such reckoning)?

Bush couldn't have imagined, in the midst of so very much that he did imagine, that information would be used mostly for itself. The users of data engage in using data. In doing so they create and become data. Data is the name of a popular television hero. Like most data, he came from the future.

Our data who art in data data be thy data, thy data come, thy data be done, on data as it is in data (to paraphrase Hemingway).

A mathematician is not a man who can readily manipulate figures; often he cannot. He is not even a man who can readily perform the transformation of equations by the use of calculus. He is primarily an individual who is skilled in the use of symbolic logic on a high plane, and especially he is a man of intuitive judgment in the choice of the manipulative processes he employs.

All else he should be able to turn over to his mechanism, just as confidently as he turns over the propelling of his car to the intricate mechanism under the hood. Only then will mathematics be practically effective in bringing the growing knowledge of atomistics to the useful solution of the advanced problems of chemistry, metallurgy, and biology. For this reason there will come more machines to handle advanced mathematics for the scientist. Some of them will be sufficiently bizarre to suit the most fastidious connoisseur of the present artifacts of civilization.

Again there is the gadgeteer's chirpy American optimism in this phrase, "the growing knowledge of atomistics. The legacy of turning over the propelling of our lives to the intricate mechanism under the hood." Some of it growing in the bones of us downwind of the Manhattan Project, Hiroshima, Chernobyl, Three Mile Island.

Even so, the same technology that prospectively dates the future—for example in the silent clock of coming death within certain irradiated living cells, now also retrospectively and routinely dates the past. My sister is an archaeologist of the Maya and like many such scientists she has occasion to use such "atomic dating" although she is as likely to date something by the drift of actual time in a given place, the wave upon wave of artifact and human remains that wash across history before us, and away from which we have bravely, if foolishly, surfed off in thinking to construct a network outside history. Bush imagines "machines to handle advanced mathematics for the scientist" that "will be sufficiently bizarre to suit the most fastidious connoisseur of the present artifacts of civilization." Among the present artifacts of civilization we used to consider the past.

The scientist, however, is not the only person who manipulates data and examines the world about him by the use of logical processes, although he sometimes preserves this appearance by adopting into the fold anyone who becomes logical, much in the manner in which a British labor leader is elevated to knighthood. Whenever logical processes of thought are employed—that is, whenever thought for a time runs along an accepted groove—there is an opportunity for the machine. Formal logic used to be a keen instrument in the hands of the teacher in his trying of students' souls. It is readily possible to construct a machine which will manipulate premises in accordance with formal logic, simply by the clever use of relay circuits. Put a set of premises into such a device and turn the crank, and it will readily pass out conclusion after conclusion, all in accordance with logical law, and with no more slips than would be expected of a keyboard adding machine.

Just when you begin to think that Van the Tinkerer is beyond our common cares, there comes a turn of phrase (these days we would immediately suspect the speechwriter) that washes over you with recognition. "Whenever thought for a time runs along an accepted groove, there is an opportunity for the machine," Bush writes. It is difficult given the prescience of this section—with its evocations of credit cards, ATM networks, target marketing, expert systems, and software agents—not to suppose that he also could have imagined the now ubiquitous machines that do not merely exploit but occupy the time when time runs along its groove: the TV tube, its brother the web, the cell phone, the vibrator, the home espresso machine.

(In their rage the SRL machines flail and thunder, spewing fire against the course of time in its groove. In the back of the closet in the hermitage of the stored-away carry-on bag stuck the onyx-cased German travel clock clicks patiently through a century.)

But Bush is really an engineer here, even the sideways nod to the logic machine means to prepare a schematic for the Memex as Gedanken Experiment. The essay itself is a logic machine: First the need to render time into a record, then the photographic interface, then the logic machine (complete with "substituting thermionic-tube switching for mechanical switching"), next the desk, tomorrow the network.

Speaking of networks, the first time I ever heard Esther Dyson (who, when I first wrote a version of this for the online webzine *Feed,* was linked in a similar enterprise parallel to whatever I was thinking and who as I wrote I could therefore imagine much like I imagine myself, floating at some outer edge of the smooth science-white surface of Bush's Leibnizian monadic surface) was at the first-ever international ACM Hypertext meeting in 1987 where we each spoke. She too imagined a logic machine, in which "good thought would easily be discriminated from bad." I will always remember this, it was the getaway keynote. I wonder if she does. I wonder is memory true. So did Bush, who suspected it mightily, perhaps at his age, fiftyish like my own now, losing his (like my own now) and longing for a Memex.

> The human mind does not work that way. It operates by association. With one item in its grasp, it snaps instantly to the next that is suggested by the association of thoughts, in accordance with some intricate web of trails carried by the cells of the brain. It has other characteristics, of course; trails that are not frequently followed are prone to fade, items are not fully permanent, memory is transitory. Yet the speed of action, the intricacy of trails, the detail of mental pictures, is awe-inspiring beyond all else in nature.

> Man cannot hope fully to duplicate this mental process artificially, but he certainly ought to be able to learn from it. In minor ways he may even improve, for his records have relative permanency. The first idea, however, to be drawn from the analogy concerns selection. Selection by association, rather than by indexing, may yet be mechanized. One cannot hope thus to equal the speed and flexibility with which the mind follows an associative trail, but it should be possible to beat the mind decisively in regard to the permanence and clarity of the items resurrected from storage.

And so finally the holy of holies for the hypertextual elite. I've quoted the "mind works by association" as much as most hypertext systems developers, hyperfiction writers, and hypertextual theorists. In fact it was interesting to speculate when I first wrote a version of this essay for an online journal, writing in isolation but knowing my links would appear along with other writers' links as encrustations upon the commented text, whether there would indeed be room enough, or any room (the problem of overlay links not having been solved for all your applets and philosophies, Horatio) to link into that oft-quoted space. It's the Potsdammer Platz of this text (a seemingly casual mention meant to summon not only a memory of a hypertextual weekend I had spent in Berlin previous to my writing this, but also of the wasteland that Bush's gadgets made of that place just months before his writing his essay).

The mind works by association. By God, yes, this is the well point, the beginning of the Hudson of hypertext, its Mecca, its Ebbetts Field, its mother's womb, the *axis mundi,* the place where you first fell in love, Dresden's Altmarkt in 1943. Faced with Bush's passing claim that while man may not "duplicate this mental process artificially . . . in minor ways he may even improve [it], for his records have relative permanency," it is surprising to me this time around to see how much this essay seems another (albeit gentler) war, this one the one we all fight, against memory and its losses.

"Relative permanency": *Ars longa, vita brevis.* As often as I quote Bush in my own writing, I am likely to quote Horace's phrase from *Ars Poetica: exegi monumentum aere perennius.* "I have built a monument more lasting than bronze" (see chapters 5 and 6 above).

"It should be possible to beat the mind decisively in regard to the permanence and clarity of the items resurrected from storage." It should be but it isn't because minds have to register this clarity and permanence. The Memex is a meditation on mortality, a summoning of the logic machine in support of the "never again" of history.

It doesn't work. The web, the Memex, the mind. We all forget and are forgotten. Next was once. This is why we have history in common and communities wherein we embody it in generations.

The owner of the memex, let us say, is interested in the origin and properties of the bow and arrow. Specifically he is studying why the short Turkish bow was apparently superior to the English long bow in the skirmishes of the Crusades. He has dozens of possibly pertinent books and articles in his memex. First he runs through an encyclopedia, finds an interesting but sketchy article, leaves it projected. Next, in a history, he finds another pertinent item, and ties the two together. Thus he goes, building a trail of many items. Occasionally he inserts a comment of his own, either linking it into the main trail or joining it by a side trail to a particular item. When it becomes evident that the elastic properties of available materials had a great deal to do with the bow, he branches off on a side trail which takes him through textbooks on elasticity and tables of physical constants. He inserts a page of longhand analysis of his own. Thus he builds a trail of his interest through the maze of materials available to him.

And his trails do not fade. Several years later, his talk with a friend turns to the queer ways in which a people resist innovations, even of vital interest. He has an example, in the fact that the outranged Europeans still failed to adopt the Turkish bow. In fact he has a trail on it. A touch brings up the code book. Tapping a few keys projects the head of the trail. A lever runs through it at will, stopping at interesting items, going off on side excursions. It is an interesting trail, pertinent to the discussion. So he sets a reproducer in action, photographs the whole trail out, and passes it to his friend for insertion in his own memex, there to be linked into the more general trail.

The language here is that of parable and it is interesting to imagine the meeting of such friends, separated by years, and yet about to be linked by a shared trail and the history of arms and how they conquer distance. I suppose it isn't necessary for the two to have been separated but, if they were not, one wonders why their conversation hasn't turned before to "the queer ways in which a people resist innovations." Could it be that the friend is among such people? He (or she, I suppose, though one wonders after the dreams of stenos) may have previously visited with Bush's character, "the owner of the memex," seen its windows, heard its ratcheting gears as the film moves into place, and yet resisted its allure.

Now the talk turns unaccountably to long bows (these friends are telling war stories of a sort, the war against the loss of memory) and it offers an opportunity to raise what they had not raised between them before. I think the visitor is named Mephistopheles and the owner of the Memex is Lola Montez (or perhaps Leni Riefenstahl). It's of course foolish of me to think so. The owner's gender has been identified through masculine pronouns. Nor would either woman have been likely to let her knowledge of "the elastic properties of available materials" lead her to branch off "on a side trail which takes [one] through textbooks on elasticity and tables of physical constants."

O inconstant world, thinks Riefenstahl (or Montez).

The trails of the owner of the Memex "do not fade." Yet he cannot have imagined the speed with which, like the surf, they now appear, fade and appear and fade again and again, the whitecaps hissing "next, next" but lapsing at the beach.

> Wholly new forms of encyclopedias will appear, ready-made with a mesh of associative trails running through them, ready to be dropped into the memex and there amplified. The lawyer has at his touch the associated opinions and decisions of his whole experience, and of the experience of friends and authorities. The patent attorney has on call the millions of issued patents, with familiar trails to every point of his client's interest. The physician, puzzled by its patient's reactions, strikes the trail established in studying an earlier similar case, and runs rapidly through analogous case histories, with side references to the classics for the pertinent anatomy and histology. The chemist, struggling with the synthesis of an organic compound, has all the chemical literature before him in his laboratory, with trails following the analogies of compounds, and side trails to their physical and chemical behavior.

> The historian, with a vast chronological account of a people, parallels it with a skip trail which stops only at the salient items, and can follow at any time contemporary trails which lead him all over civilization at a particular epoch. There is a new profession of trail blazers, those who find delight in the task of establishing useful trails through the enormous mass of the common record. The inheritance from the master becomes,

not only his additions to the world's record, but for his disciples the entire scaffolding by which they were erected.

This last section is another of the holy texts of hyperdom and I, too, have genuflected here enough at the shrine of the "new profession of trail blazers," albeit without fully noticing quite how they serve "the master" whose "inheritance . . . becomes . . . not only his additions to the world's record, but for his disciples the entire scaffolding by which they were erected."

Among this curious list, historian, physician, chemist, and lawyer, the patent attorney stands out. Trailblazer for the engineer, attentive to his client's interests, scaffold and addition alike.

This time, however, I am caught by yet another shrine, this one again devoted to the mysterious body of the dreaming steno.

We know that when the eye sees, all the consequent information is transmitted to the brain by means of electrical vibrations in the channel of the optic nerve. This is an exact analogy with the electrical vibrations which occur in the cable of a television set: they convey the picture from the photocells which see it to the radio transmitter from which it is broadcast. We know further that if we can approach that cable with the proper instruments, we do not need to touch it; we can pick up those vibrations by electrical induction and thus discover and reproduce the scene which is being transmitted, just as a telephone wire may be tapped for its message.

The impulses which flow in the arm nerves of a typist convey to her fingers the translated information which reaches her eye or ear, in order that the fingers may be caused to strike the proper keys. Might not these currents be intercepted, either in the original form in which information is conveyed to the brain, or in the marvelously metamorphosed form in which they then proceed to the hand?

Bush imagines a technology capable of capturing "the impulses which flow in the arm nerves of a typist [and] convey to her fingers the translated information which reaches her eye or ear." This is a new instrumentality of the kind through which

Bush says "science may implement the ways in which man produces, stores, and consults the record of the race."

Bush, whom a recent book credits with implementing the profession of engineer, sees this implementation in engineering terms. The Memex is an older machine than we have ever imagined it to be, one that transforms "the marvelously metamorphosed form" of the body (in this case a woman's body in fact) into another trail to blaze, another addition to the world's record.

A Part: *Honi soit qui mal y pense*

He often sang "O Susanna." Sometimes he played the harmonica. I never thought him a technologist, although by the time of his death he had mastered Atari BASIC well enough to send a railway train of graphics sprites chugging across a screen, the flatcars carrying a load of words, "T-O-M I-S A J-E-R-K," the letters of which it dumped as it derailed, the whole thing accompanied by vaguely Asian music and Mario Brothers sound effects, the chug, the whistle, the catcall, clatter and laughter as the letters tumbled from the wireframe flatcars.

Tom is my brother's name and was my father's as well. There was no irony here, no mistaken identity, no psych-eye-a-tree as he always phrased it (be-you-tee-full also given all its syllables). The intended audience was clear. My father would turn red with laughing and my brother tried to be jovial and earnest. There was no enmity between them. In my father's last years my brother made it his project to know him. My father's laughter, like my own, could sometimes reach the high pitch of the Atari laughter.

The technologies he mastered in his lifetime include the various cameras of a professional (Hasselblad, Rolleiflex, and Grafflex); the furnaces of a steelworker (at the end of his working life he had risen to become a melter, the ad hoc metallurgist and chemist responsible for steering 250 tons of molten steel through the narrow channel of a spec), including the blast furnace, the open hearth, and the BOF, basic oxygen furnace, the one that replaced steelmen with college boys and eventually closed the plant; also the various forms of sound and image transmission, from shortwave radio to television, some of which he built from kits, some that he repaired with the ever-present volt-ohm meter and tube tester; various modes of transportation, the car, the truck, the car and camper trailer, the narrow-gauge railway engine, the bulldozer, the earthmover, motor boats, and the eighteen-foot-long high-prowed wooden Great Lakes rowboats from which he used to fish; the technologies of fishing themselves, mostly the braided drop-line with lead sinkers or the ice fishing tip-up or the long bamboo poles without reels—he never really liked the spinning rod; camp gear of all sorts, tents, dining flies, pop-up campers, a small trailer, Coleman stoves and lanterns, huge coolers, carving knife (he could whittle a fetish woman from a fat twig), axe, machete (he nearly severed an artery, chopping wood while camping with my mother and youngest brother, stayed long enough to calm everyone and then drove alone to the hospital, steering with the wounded—now tourniqueted—arm, holding a compress against the wound); he taught himself the piano, seemed to know how to play the concertina (called a squeeze box in some vaguely Irish-American way), didn't dance, sang like a melancholy god. And, oh yes, the technologies of literacy: he knew and read Latin and Greek, described himself as "Jesuitical . . ." and after a pause, "self-taught," in the last years of his life kept a journal in leather-bound American Express daybooks in which he marked the weather, illness, debts, visits of my siblings and in which he referred to himself as "self," that is, "self tired, grey day." His grandfather on his mother's side was fabled for lecturing in English while writing on a chalkboard in two other languages, one with each hand; he himself loved a book called *The Loom of Language.* He had the usual Irish store of songs and poems that the eldest son I do not have. Learned, as noted, BASIC, was impressed by the notion of a business language like COBOL, thought LISP sounded interesting. Loved the mysteries of sine, cosine, and tangent, was a master of the slide rule even into my high schools days when he taught me to do equations on an elaborate one with several scales; easily grasped the graphing calculator; claimed to know the abacus, however his only card game pinochle.

I don't know if he rode a bicycle, for a period he did ride to

work at the plant clinging to a coworker on the back of a motor-
cycle until a guy they knew flipped off his Indian cycle at high
speed and was decapitated by overhead telephone wires. Despite
long study he never became a ham radio operator though he
knew Morse code. He devised a method of keeping his checkbook
on the Atari but it required that balances be listed as negative
figures and overdrafts as positive. Did not play golf, tennis, baseball, hockey, or football, though he earned a letter in soccer during his year or so at teachers college, way ahead of the curve; and
he used to dive from the long skiffs to the bottom of Lake Erie
when he rowed out to fish (a scar on his palm marked where he
had pushed off from jagged wreckage on the bottom where the
perch cruised). There and elsewhere afterwards he swam out so far
you could hardly see him, floating at ease. Started at the plant at
age sixteen, continued there through the short college stint, met
and married my mother, was as I've said 4F for the war because of
a bad back, though he was strong enough to carry the small white
casket of his firstborn, my elder sister Mary, born dead, her sole
mourner and cortege. He met my mother that year at college. The
family story was that he was supposed to go to MIT after high
school, that a great uncle had offered to pay his way through, but
his parents declined because the great uncle was a scandal, having
photographed his wife in her coffin, which was thought to be a
sacrilege for reasons only Walter Benjamin could explain.

Bush would still have been there at MIT in 1934, vice president and dean, working on optical and photocomposition
devices as well as a machine for rapid selection from banks of
microfilm, when my father was eighteen.

It strikes me that one thing we never questioned in the story
of my father's lost chance for MIT was that he would have been
accepted there. I can't recall when I truly came to understand
class structure, although it was comparatively late in life, long
after a childhood in a union family, my adolescent (and continuing) social (and quasi-socialist) activism, my iconoclast pedagogy
and aesthetics, my meager technologies and their minor innovations.

"Can I help you, hon?" the steno asks my father, her red nails
like Chinese lacquer.

Now this is, of course, questionable. Despite her own class differences we can assume that the steno must have enough finish about her to have seemed suitable for the socially awkward Dean B. Perhaps she said "son" not "hon." Perhaps my father is thinking of the woman named Tootsie who ran in their circle (as they said), a steno whose rhinestone glasses and red, red lips were crowned by a flounce of honey blond hair.

Though he hasn't met her yet, would not have probably, had he gone off to Cambridge while his younger brothers grew and ran in circles with babes with red, red lips and dark, or honey or auburn or red, red hair and went away to war.

Toot toot Tootsie good-bye.

(Oh the women of those pictures then—the photographs largely my father's—are sexy even at this distance, full hipped and buxom with laughing eyes and stories enough in my slipping memory to make it clear they were no bimbos. They carried families through the war and worked at important things. Still they were sexy, sexier for all this, however strange it may seem to suggest that desire stirs before the photographic images of one's aunts and their friends. Even so my brothers and I have wives, and any of us can remember the body of the girl pressed up against the sailor in Times Square following VJ Day in that famous picture, holding her hat, her foot kicked up behind her as he dips her down deep into the kiss.)

"I can come back," my father almost surely would have said, longing for the infinite hallway, all his life reticent about women, despite the eight (nine, counting my sister before me whom he carried to her grave) children, though charming them nonetheless, from Sears clerks to long-distance operators, the blue eyes of my father.

I think I learned from him to assume that I do not belong. Rule-bound though anarchic, caste and class sensitive though transgressive. Always out of place. (Do you think this whole narrative should have been woven through the meditation upon Bush's text? There's a kind of modernist reassurance to that kind of strategy, making it seemingly more suitable for reverie and experiment. The mere appendage of speculative history is a naive textual strategy. The way stories are told.)

How do we learn any technology, writing for instance? The bulldozer?

I remember as a boy somehow knowing that the jokes my aunts made were bawdy, even when I didn't understand the references, from the tone as they told them, their naughty laughter and the wonderful way they moved, pale arms sometimes swathed in sleeves of see-through netting, red satin flowers on their tailored jackets.

"Did you want to see the dean?" the steno asks. The impulses that flow in the arm nerves of a typist convey to her fingers the translated information that reaches her eye or ear, in order that the fingers may be caused to strike the proper keys.

My father loves her. Anyone would.

He does want to see the dean but he doesn't know what he wants to know. I am shocked to see how much my own sons know of college, whether from within it with my eldest or from the prospect of his younger brother, a high school senior. Taxonomies and categories of knowledge were a mystery to me and so I am badly educated, even now, though this makes me a good teacher, someone known for breaking bounds, not knowing anything else.

The dean's door is slightly open and there is the machine, the Memex, miraculously built a decade before its invention, also before the invention of the terrible engines and furnaces that threw the physicists so violently off stride. A table of light, two side-by-side windows, the soft whirring of the geared beds that search through and serve up the microfilm. Human knowledge shines forth to my father from the inner office, the scientist's face softly illuminated by the backlight. The curve of the longbow.

The steno waits to see what my father will do. The lady or the tiger.

> Consider a future device of individual use, which is a sort of mechanized private file and library. It needs a name, and to coin one at random, "memex" will do. A memex is a device in which an individual stores all his books, records, and communications, and which is mechanized so that it may be consulted with exceeding speed and flexibility. It is an enlarged intimate supplement to his memory.

She needs a name, and to coin one at random, Beatrice will do. (How do we learn any technology, writing for instance?)

My father has never been so far away from home. Sometimes in summers they all went back down to Pennsylvania, periodically reversing the migratory flow which brought them north (my mother's family also, perhaps no coincidence, though they didn't know each other, hers from Wilkes-Barre, his Duryea, both outside Scranton—and Beatrice? she is from Roxbury, Irish Boston), having to drive up certain hills (the Warsaw Hill in New York, nameless hills in Pennsylvania) in reverse to gain a low enough gear to conquer the grade.

The talk turns to the queer ways in which a people resist innovations, even of vital interest.

Speaking of gear ratios, Dr. Bush displays the diagram of a winch upon the screen of the Memex.

"It could lift a whole steamship," he says. "Guess the century?"

The girl strokes the keys languidly and looks about the room and sometimes at the speaker with a disquieting gaze.

No guess here. My father will be wrong in any case, and I am not well versed enough in the history of science to suggest an anomalous enough century.

On his mother's side, my father's family stayed there, in Mount Pleasant, Pennsylvania, and vicinity. Education had made them nearly gentry. "The Professor," as they always called his grandfather, had become superintendent of schools.

My father's life moved from union to management, forty-five years in the plant: laborer, helper, first helper, foreman, turn foreman, melter. Even when he became a foreman he felt out of place, not unlike Bush's British labor leader elevated, by logic, to knighthood.

It has other characteristics, of course; trails that are not frequently followed are prone to fade, items are not fully permanent, memory is transitory. Yet the speed of action, the intricacy of trails, the detail of mental pictures, is awe-inspiring beyond all else in nature.

And still we know nothing of the stenotypist. Is this a failure of imagination? Or is the story of this particular technology still so suffused with gender that it becomes as translucent as greased paper.

"Have you studied optics?" Dr. Bush asks him.

Beatrice O'Leary late of Roxbury wears herring boxes without topses instead of shoes. (He sang this song also, O my darling Clementine) She sells seashells by the seashore. My great-great uncle photographed her in her coffin, alabaster.

Under the right circumstances my father could have participated in a great experiment that threatened to end all creation. Under similar circumstances he might have developed the Cyclops camera or the hypertext novel. There isn't a genealogy of technology, only a series of accidental terrains. (Dr. Bush's masters thesis included the invention of the Profile Tracer, a surveying tool meant to measure distances over uneven ground; my father died before the popularity of fractal mathematics that he would have loved, in love with cosines.)

Selection by association, rather than by indexing, may yet be mechanized.

What are the choices? To make her motherly and maternal, mom to all the boys who come to see Dean B.? Make her sirenlike, with China red nails and perfumes, the women who might have been my mother had my father's uncle not loved his wife so well. Hon or son? Have her chew gum, have her read Proust, not have her at all but let her be as she may.

Surely there are some readers who have left us long ago, seeking sex or wisdom, weary of elegiac word games and, however seductive, indirection. I should quote something else, always an interesting gambit, although I know less and less as I go on, quote the single citation here only in its electronic evanescence, sans pages, without the smell of old newsprint from postwar magazines or the chemical fume of the glossy magazines of my youth, *Colliers* and *Life,* the latter where Bush's essay was reprinted the following summer with pictures of the—virtual—Memex:

> Memex in use is shown here. On one transparent screen the operator of the future writes notes and commentary dealing with reference material which is projected on the screen at left. Insertion of the proper code symbols at the bottom of right-

hand screen will tie the new item to the earlier one after notes are photographed on supermicrofilm.

On the screens in the illustration: longbows.

Consider the orientation of one screen to another as an instance of some undetermined temporal relation. The screen on the left seems to take the place of memory, the screen at the right consciousness or what we call the present. The stenotypist sits in similar relation to the office within. She is the present tense body, Doctor Bush is embodied memory, a walking time machine.

In his pornographic gaze he imagines intercepting the impulses between her eye or ear and her finger. More contemporary feminist critics have rightly criticized the similar phallo-networkism of Gibson's protagonist "jacking in" to cyberspace.

My father does not know what to do.

Consider the orientation of one discourse to another, fiction to informal exegesis, as an instance of some undetermined spatiotemporal relation.

Let us return to his photographs of my aunts and their circle again, at this point in the imaginary future perhaps including Beatrice instead of my mother, though we could suppose the other aunts will remain the same. Given what we now know of the male gaze, what can we say of erotic feelings awakened by photographs of a dead woman (you could think of Marlene Dietrich or French postcards from the turn of the century if it is awkward to think of my aunts and their gang, the other term they used for their circle)?

When we were small we carried holycards with a buxom virgin, in the best ones her head was wreathed with actual gold, or so we thought it.

What is the relation of the body to memory?

I would urge my father to ask this same question of Dean Bush but I don't want to imperil his chances at MIT since it has taken him such struggles to get here. In his jacket pocket there is a cardboard folder with the pictures of his great aunt. Alabaster, she seems merely sleeping and the thought of the sleeping woman arouses his sensibilities. From somewhere comes a dim memory of the dark perfume of a woman asleep. It is my mother.

What is the relation of image to text when each is in motion?

The stenotypist has the fragrance of lilac soap. She looks about the room and sometimes at him with a disquieting gaze.

I am being born. Insertion of the proper code symbols at the bottom of right-hand screen will tie the new item to the earlier one after notes are photographed on supermicrofilm.

"Yes, I've seen the future," my father answers. "The camera never lies."

I am rather proud of him for this. He means a careful statement regarding mediation. The camera is an instrumentality, a transmissive machine, and so is without any moral dimension or truth value. Of the veridicality of the supposed world before it, my father is not so certain. The truth nature of the film seems even more questionable, something like what we Catholics believed of the Communion Host and called by the name transubstantiation. The true body of Christ. Next was once.

Bush can hardly hide his displeasure. There are many special machines, such as the harmonic synthesizer that predicts the tides.

"Poetry," he says, without emotion, one way or another.

It tells us nothing.

The stenotypist strokes the keys of the Voder and it emits recognizable speech.

"S' io ridessi," mi cominciò, "tu ti faresti quale fu Semelè quando di cener fessi . . ."

Were I to smile, she told him, you'd turn to ash as sure as Semele.

No human vocal cords entered in the procedure at any point; the keys simply combined some electrically produced vibrations and passed these on to a loudspeaker.

"Falling in love again, never wanted to, what am I to do, I can't help it."

On one transparent screen the operator of the future writes notes and commentary dealing with the video of Dietrich singing that is projected on the screen at left.

Though my father knows little of Dietrich or Dante, he has studied Greek mythology and knows the nine names of the moon, among them Semele, mother of Dionysus, "the child of the double door," he who was born of Zeus's thigh after Semele was consumed by his fire and lightning, misled by Hera into asking to experience his true nature. It was Hermes, the god of crossroads (I wrote crosswords) who devised the birthing technology. This kind of knowledge is passing like the Turkish bow. Perhaps the war against memory is a struggle against embodiment and birth itself, the double portal of memory.

The way these time travel narratives work is a knot.

This was the aim of the experiments: to send emissaries into Time, to summon the Past and Future to the aid of the Present.

But the human mind balked at the idea. To wake up in another age meant to be born again as an adult. The shock would be too great.

Having only sent lifeless or insentient bodies through different zones of Time, the inventors were now concentrating on men given to very strong mental images. If they were able to conceive or dream another time, perhaps they would be able to live in it. (Marker 1992, 7)

The risks are known. Even so I choose to let my father stay there, hesitating in the space between the two doors, the double screens, between the father of hypertext and the steno whose impulses were interrupted by a machine. What are the possibilities? That I will not be born, that time will go on without me.

It strikes me that in much the same way that we never questioned his admissibility in the story of my father's lost chance for MIT, my first inclination is not to think that, left there, he might have affected the face of history in the way Bush did. More: that he might have affected Bush's own subsequent history and thus, to the extent that my life and thinking has been shaped by the Memex, my own.

It is more so telling that I do not imagine this scene with my mother entering history, she the reticent student at the dean's door (or did you think her the steno?).

Time is a slip knot.

The stenotypist's is the face of history, like Benjamin's angel looking backward.

> This time he is close to her, he speaks to her. She welcomes him without surprise. They are without memories, without plans. Time builds itself painlessly around them. Their only land-marks are the flavor of the moment they are living and the markings on the walls.

> Later on, they are in a garden. He remembers there were gardens. (Marker 1992, 17)

The lady or the tiger. Was or next. I can be supposed to wish that my father would choose to speak to the man within the office, although all my life my own inclinations have been other-wise. If I thus seem to have set up a—however badly dramatized—polarity, there in fact is none. Rather consider this an instance of some undetermined temporal relation. Given only these choices I would he chose mystery. Given life to do over again, I would have him choose as he did, not so that (or not merely so) I might become my own mother's son but so that the possibilities that he here encounters in my imagination would continue to have the same retrospective richness of potentiation.

You are correct if you see in this something of an allegory of hypertextuality.

I am sorry if this seems too clever by far or not clever enough for you. It is my life story and can only take this form. This is how it ends so far.

Portrait of the Artist as a Search Engine Entry

The most likely portrait of you that would emerge if you got run over by a laundry truck wouldn't come from the internet but from the contents of your wallet. Although that probably won't be true for long. Already half of what is in my wallet are computer interfaces. In fact if you want to understand what people mean when they talk about the shift from a print to a digital age, look at your credit cards. The front with the bumps for your name and account number is print, the back with the electromagnetic tape is digital. Just think about how long it has been since somebody mashed a roller over the bumps to make you a hand-printed copy of your life as a consumer.

The ATM and credit card stripes that Amex and ShopRite et alia use to compile information about me could just as easily tell an ambulance attendant my blood type, recent shopping list, and list of publications (with graded reviews). In the future envisioned by Nicolas and Newt alike, the ambulance guy could do his own brand of complete assets search. He could have a look at my HIV test results, credit history, and Pennysaver notices and decide on the spot, so to speak, whether I'm a deadbeat or deadmeat and literally haul my ass off to the genome recycler.

Even so, for the moment at least it still makes some sense to ask which one tells more about you, the contents of your wallet or the trail you leave as you pass like a comet through cyberspace. I got thinking about all this recently when in the midst of shutting down a half-dozen computers my students had left open, I found a screenful of hits from a search engine with my name on them. Someone (not me, I swear, though I can't plead eternal innocence) had whiled away class time looking up links to my bona fides.

Where am I more there, I wondered, on the web or in my wallet? After having a brief look at each, I suppose it depends on what you want to say about yourself and whom you want to say it to. When your life flashes before your eyes on the computer screen, you begin to have these mortal memories of Laminated Sanity and Three-Card Mommy.

Long ago when I first started teaching I used to know a guy named Ben who had his life laminated. A Yalie gone to seed, vaguely fiftyish and moonfaced as a movie friar, he had the Clorox scent, chewed raw cuticles, and deer-in-the headlight eyes of a former mental patient. His khakis were ketchup stained and his Bass weejuns out at the stitches (though he regularly replenished them with shiny copper pennies). He carried his laminated life in the inside pocket of a natty, nasty Brooks Brothers olive tweed jacket, unpressed, fragrant, and frayed.

"Look here, Mr. Joyce," he would say, fanning the yellowing documents out in stubby, tobacco-yellowed fingers. I was his teacher then at a community college in Michigan where we each had wandered far from the great white way. "I used to be a reporter for the *New York Times* and I am one of few people you know who can actually prove his sanity."

He probably should have laughed at the last but he occupied a life far from laughing at himself, in a place where no one liked irony because (as the poet Charles Olson puts it) "the iron in it confused them."

There were, I think, three or four short pieces in his laminated portable clipbook, at least one with his by-line. Ben was happy that I could see the pieces were from the *Times* by their typestyle. In the center of the fan, like a three-card monte dealer's tease, was a much larger document: his release from the mental hospital. I don't think it used the word "cured" exactly. It probably said something about "successful functioning." He was the first person I knew who held out the promise of holding your life in your hand.

There isn't a lot in either my wallet or on the web that definitely points to my sanity or even my successful functioning. Three things give the edge to my wallet at the moment. I suppose I could call them pieces of information, I suppose I *should* call them documents. The two wallet-size school pictures of my kids lack the wise-ass, fatally hip, sentimental, or slightly smutty inscriptions that high school pals provide each other. Still something of the mischief in their eyes and the moody poseur attitudes provide a portrait of me as much as them. As does the fact that I have these images here with me. It's something you can't find out about me on the web.

But that is my fault not the web's. In fact, in the course of making my way link by link through remote sites on the day that I surfed the web looking over the comet's trail of evidence of myself as seen through a search engine, I was distracted for nearly an hour first by family snapshots on the homepage of an English major at a college in Oregon, and then by full-color scans in the "art gallery" on the homepage of a student poet and physicist and painter at a certain northeast liberal arts college. The Oregon snapshots were like the wallet-size snapshots from friends that my sons sometimes leave around: an intimate glimpse into school life and budding notions of romance and poetry. In this case, however, the snapshots were left out for the whole world to see on the web. The northeast paintings were lively, figurative (one was a touchingly unsmutty nude self-portrait), painterly even in their digitized reproduction, though most likely what an art critic would find naively pseudoimpressionist and mannered. I thought they were pretty honest with emphasis on each of the two words: pretty and honest.

What was missing from the search engine portrait of myself as literary-technological dimly semicelebrity was neither the adolescent romance and poetry of the Oregon student nor the pretty honest impressionism of the northeast student. What was missing was the physicality of their presence there. Not just the physical images of snapshot and self-portrait but the physical sense that they left parts of their lives strewn there like the contents of a wallet or a desk drawer. What was missing from my cyber-portrait was the naive mix of coherence and happenstance left out for a world to see. Their homepages were animated not laminated.

It is this animated mixture of coherence and happenstance that at least once promised to make the web a brave new world, if not for me at least—as Shakespeare's poet and physicist and multimedia celebrity father-figure, Prospero, said of his own teenage romantic daughter, Miranda—to you.

Seeing the student homepages, like any father, I wanted another chance at life.

For the other thing my wallet has that my search engine portrait doesn't is palpable history, the embodied second chance. And this will be harder to come by even on the most animated homepages. In this case I mean history as document: a forty-year-old, worn business card for Thomas R. Joyce, Jr., his name centered in a stylish rotogravure display font above the title "Photographer" and under that, on two lines, "Weddings—Home Portraits/Commercial." The four corners of the card are carefully balanced with symmetrical type slugs. On the bottom left the street address, the bottom right the town. On the top left a lost kind of phone number "Derby 4196," and on the top right a lost kind of caring: the miniature logo for the Printers Union Local 14. My father took photographs to get us through strikes at the steel plant where he worked as a union man.

Nothing I carry in cyberspace corresponds to what I carry with this card in my wallet. Perhaps nothing can. Still even outside the web there are limits to how much life you can hold in your hand. This too is a history I carry with me though in no form other than stories.

In one story my father and I sit in Boomer Cannan's funeral parlor in the Irish ghetto of South Buffalo, New York, settling the details of my mother's funeral.

"Oh, Tommy," the Boomer says to my father (they worked together in the steel plant in real life). "She looks like hell after this cancer. It's better this way." Then he steps out of the room to look for some sample holy cards for us to choose from.

My father looks up at me and holds out a sheaf of documents not unlike Ben's laminated history or my search engine portrait. "Forty years of marriage," he says, "and all you have left is a hand full of papers."

Later on, after his death, we found that he had saved her last batch of Christmas cookies in a desk drawer. A few years after that I sprayed plastic fixative over one of them, a red sugar wreath with gold candy stars, and then matted and framed it to preserve it for myself. I'm thinking now that I might just scan both the cookie and the card for my boys and then put them both up on the web.

Bibliography

Aarseth, Espen. 1997. *Cybertext: Perspectives on Ergodic Literature*. Baltimore: Johns Hopkins University Press.

Amerika, Mark. 1997. *Grammatron*. <http://www.grammatron.com/>.

Appadurai, Arjun. 1996. *Modernity at Large: Cultural Dimensions of Globalization*. Minneapolis: University of Minnesota Press.

Armstrong, Cara, and Karen Nelson. 1993. "Ritual and Monument." *Architronic*. <http://www.saed.kent.edu/architronic/v2n2 /v2n2.05 .html>.

Arnold, Kenneth. 1994. "The Electronic Librarian Is a Verb/the Electronic Library Is Not a Sentence." Gilbert A. Cam Memorial Lecture, New York Public Library. *Journal of Electronic Publishing*. <http:// www.press.umich.edu/jep/works/arnold.eleclib.html>.

Arnold, Mary Kim. 1993. "Lust." Storyspace software for Macintosh and Windows. *Eastgate Quarterly Review of Hypertext* 1, no. 2.

Bachelard, Gaston. 1964. *The Poetics of Space*. Trans. Maria Jolas. New York: Orion Press.

Balpe, Jean-Pierre. 1991. *L'Imaginaire informatique de la littérature*. Paris: Presses Universitaires de Vincennes.

Bazin, Patrick. 1996. "Toward Metareading." In *The Future of the Book*, ed. Geoffrey Nunberg and Patricia Violi. Berkeley and Los Angeles: University of California Press.

Beckett, Samuel. 1960. *Krapp's Last Tape and Other Dramatic Pieces*. New York: Grove Press.

———. 1995. *The Complete Short Prose: 1929–1989*. Ed. S. E. Gontarski. New York: Grove Press.

Bennahum, David. 1994. "Fly Me to the MOO: Adventures in Textual Reality." *Lingua Franca,* May–June 1994, 29–31.

Bernstein, Charles. 1986. *Content's Dream*. Los Angeles: Sun and Moon.

Birkerts, Sven, ed. 1996. *Tolstoy's Dictaphone: Machines and the Muse at the Millennium*. St. Paul: Graywolf Press.

Birkerts, Sven, Carolyn Guyer, Michael Joyce, and Bob Stein. 1995. "Page versus Pixel: The Cultural Consequences of Electronic Text." *Feed,* June. <http://www.feedmag.com/95.05dialog1 .html>.

Bolter, Jay David. 1984. *Turing's Man: Western Culture in the Computer Age*. Chapel Hill: University of North Carolina Press.

———. 1991. *Writing Space: The Computer, Hypertext, and the History of Writing*. Hillsdale, N.J.: Lawrence Erlbaum and Associates.

Bolter, Jay, and Richard Grusin. 1998. *Remediation: Understanding New Media*. Cambridge: MIT Press.

Bootz, Philippe. 1994a. "Video Poem and Unique Reading Poem: Two Complementary Approaches, Bi-lingual French-English." In

A:\LITTÉRATURE, colloque Nord Poésie et Ordinateur. Lille, France: CIRCAV-GERICO and MORS-VOIR.

———. 1994b. *Hommage à Jean Tardieu.* Bound-in disk for Macintosh and Windows in *A:\LITTÉRATURE, colloque Nord Poésie et Ordinateur.* Lille, France: CIRCAV-GERICO and MORS-VOIR.

———. 1994c. *Passage (extract) Unique Reading Poem.* Bound-in disk for Macintosh and Windows in *A:\LITTÉRATURE, colloque Nord Poésie et Ordinateur.* Lille, France: CIRCAV-GERICO and MORS-VOIR.

———. 1996. "Poetic Machinations." *Visible Languages* (Rhode Island School of Design) 30, no. 2: 118–38.

Borges, Jorge. 1971. *The Aleph and Other Stories: 1933–1969.* New York: Bantam Books.

Braman, Sandra. 1994. "The Autopoietic State: Communication and Democratic Potential in the Net." *Journal of the American Society of Information Scientists* 45, no. 6: 358–67.

Bush, Vannevar. 1945. "As We May Think." Prepared by Denys Duchier, April 1994. <http://www.isg.sfu.ca/~duchier/misc/vbush/>. Originally published in *Atlantic Monthly,* July 1945, 101–8.

Calvino, Italo. 1986. "Cybernetics and Ghosts." In *The Uses of Literature.* New York: Harcourt Brace Jovanovich.

Chaitkin, Samantha. 1996. Hypertext essay, Vassar College.

Cixous, Hélène. 1991. *Coming to Writing and Other Essays.* Ed. Deborah Jenson. Trans. Sarah Cornell, Deborah Jenson, Ann Liddle, and Susan Sellers. Cambridge: Harvard University Press.

———. 1993. *Three Steps on the Ladder of Writing.* Trans. Sarah Cornell and Susan Summers. New York: Columbia University Press.

Coover, Robert. 1983. "An Interview with Robert Coover." Interview by Larry McCaffery. In *Anything Can Happen: Interviews with Contemporary American Novelists,* by Tom LeClair and Larry McCaffery. Urbana: University of Illinois Press.

———. 1993. "Hyperfiction: Novels for Computer." *New York Times Book Review,* August 29, 1 and 8–11.

Cosgrove, D. E., and S. J. Daniels, eds. 1988. *The Iconography of Landscape: Essays on the Symbolic Representation, Design, and Use of Past Environments.* Cambridge: Cambridge University Press.

Coverley, M. D. [Marjorie Luesebrink]. 1997. *Lacemaker.* <http://gnv.fdt.net/~christys/elys_1.html>.

Davis, Erik. 1994. "It's a MUD, MUD, MUD, MUD World." *Village Voice,* February 22, 42–43.

Debray, Regis. 1996. "The Book as Symbolic Object." In *The Future of the Book,* ed. Geoffrey Nunberg and Patricia Violi. Berkeley and Los Angeles: University of California Press.

de Certeau, Michel. 1983. *The Practice of Everyday Life.* Trans. Steven Randall. Berkeley and Los Angeles: University of California Press.

Deleuze, Gilles, and Félix Guattari. 1983a. "The Smooth and the Striated." In *A Thousand Plateaus: Capitalism and Schizophrenia,* trans. Brian Massumi. Minneapolis: University of Minnesota Press.

———. 1983b. "Treatise on Nomadology—the War Machine." In *A Thousand Plateaus: Capitalism and Schizophrenia,* trans. Brian Massumi. Minneapolis: University of Minnesota Press.

Derrida, Jacques. 1985. "The Question of Style." Trans. Ruben Berezdivin. In *The New Nietzsche,* ed. David B. Allison. Cambridge: MIT Press.

Dibbell, Julian. 1993. "A Rape in Cyberspace: How an Evil Clown, a Haitian Trickster Spirit, Two Wizards, and a Cast of Dozens Turned a Database into a Society." *Village Voice,* December 21, 36–42.

Dieberger, Andreas. 1996. "Browsing the WWW by Interacting with a Technical Virtual Environment—a Framework for Experimenting with Navigational Metaphors." In *Hypertext '96: The Seventh ACM Conference on Hypertext.* New York: ACM Press. Also available at <http://www.lcc.gatech.edu/faculty/dieberger/ht96.paper.html>.

Douglas, Jane Yellowlees. 1991. "Understanding the Act of Reading: The *WOE* Beginner's Guide to Dissection." *Writing on the Edge* 2, no. 2: 112–26.

Dreyblatt, Arnold. 1997a. *Memory Arena.* <http://www.uni-lueneburg.de/memory/>.

———. 1997b. *Who's Who in Central and East Europe, 1933.* <http://www.uni-lueneburg.de/memory/>.

Eco, Umberto. 1989. *The Open Work.* Trans. Anna Cancogni. Cambridge: Harvard University Press.

———. 1994. "La bustina di Minerva." *Espresso,* September 30, 1994, 56.

———. 1996. Afterword to *The Future of the Book,* ed. Geoffrey Nunberg and Patricia Violi. Berkeley and Los Angeles: University of California Press.

Eliot, T. S. 1943. *Four Quartets.* New York: Harcourt, Brace and World.

Fenollosa, Ernest. 1967. *The Chinese Written Character as a Medium of Poetry.* Ed. Ezra Pound. San Francisco: City Lights.

Fleischmann, Monika. 1997a. *Liquid Views: Virtual Mirror of Narcissus.* Interactive computer video installation. <http://viswiz.gmd.de/IMF/liquid.html>.

———. 1997b. *Rigid Waves: Narcissus and Echo.* Interactive computer video installation. <http://viswiz.gmd.de/IMF/rigid.html>.

Florin, Fabrice. 1990. "Information Landscapes." In *Learning with Interactive Multimedia,* ed. S. Ambron and K. Hooper. Bellingham, Wash.: Microsoft Press.

Foster, Donald. 1996. "A Funeral Elegy W[illiam] S[hakespeare]'s 'Best Speaking Witnesses.'" Followed by the text of *A Funeral Elegy.* PMLA 111, no.5: 1080–1106.

Foster, Donald, and Jacob Weisberg. 1996. "Primary Culprit." *New York,* February 26, 50–58.

Foucault, Michel. 1967. "Of Other Spaces." *Diacritics* 16, no. 1: 22–27.

———. 1980. *Power/Knowledge: Selected Interviews and Other Writings,*

1972–1977. Ed. Colin Gordon, trans. Colin Gordon et al. New York: Pantheon.

Fulton, Alice. 1996. "Screens: An Alchemical Scrapbook." In *Tolstoy's Dictaphone: Machines and the Muse at the Millennium,* ed. Sven Birkerts. St. Paul: Graywolf Press.

Furuta, Richard, and Catherine C. Marshall. 1996. "Genre as Reflection of Technology in the World-Wide Web." *Hypermedia Design: Proceedings of the International Workshop on Hypermedia Design,* ed. S. Fraïssé et al. London: Springer.

Galef, David, ed. 1998. *Second Thoughts: A Focus on Rereading.* Detroit: Wayne State University Press.

Gibson, Stephanie, and Lance Strate, eds. 1996. *The Emerging Cyber-Culture: Literacy, Paradigm, and Paradox.* Cresskill, N.J.: Hampton Press.

Grimes, Deirdre. 1999. "A Journal for My Child." In *Mothermillennia.* <http://mothermillennia.org>.

Guyer, Carolyn. 1992a. "Buzz-Daze Jazz and the Quotidian Stream." Paper presented to the Annual Meeting of the Modern Language Association, New York.

———. 1992b. *Quibbling.* Storyspace software for Macintosh and Windows. Cambridge, Mass.: Eastgate Systems.

———. 1996a. "Along the Estuary." In *Tolstoy's Dictaphone: Machines and the Muse at the Millennium,* ed. Sven Birkerts. St. Paul: Graywolf Press.

———. 1996b. "Following Buzz-Daze." In *Interface Three: Conference Documentation,* ed. Klaus Peter Dencker. Hamburg: Kulturbehörde.

Guyer, Carolyn, Rosemary Joyce, and Michael Joyce. 1999. *Sister Stories.* <http://www.eastgate.com>.

Habermas, Jürgen. 1992. "The Normative Content of Modernity." In *The Philosophical Discourse of Modernity: Twelve Lectures.* Cambridge: MIT Press.

Haraway, Donna. 1991. "Situated Knowledges: The Science Question in Feminism and the Privilege of Partial Perspective." In *Simians, Cyborgs, and Women.* New York: Routledge.

Harpold, Terry 1991. "Contingencies of the Hypertext Link." *Writing on the Edge* 2, no. 2: 126–37. Also available at <http://www.lcc.gatech.edu/faculty/harpold/papers/contingencies/index.html>.

———. 1994. "Conclusions." In *Hyper/Text/Theory,* ed. George Landow. Baltimore: Johns Hopkins University Press.

———. 1996. Author's Note to "Contingencies of the Hypertext Link." <http://www.lcc.gatech.edu/faculty/harpold/papers/contingencies/index.html>.

Harvey, David. 1993. "From Space to Place and Back Again: Reflections on the Condition of Postmodernity." In *Mapping the Futures: Local Cultures, Global Change.* London: Routledge.

Hayles, N. Katherine. 1992. "Gender Encoding in Fluid Mechanics: Masculine Channels and Feminine Flows." *Differences* 4, no. 2.

————. 1993. "Virtual Bodies and Flickering Signifiers." *October* 66:69–91.

————. 1997. "Introduction: Situating Narrative in an Ecology of New Media." *Media Fiction Studies* 43, no. 3 (September–October): 1–5.

————. 1998. "Byte lit." *Artforum* 37, no. 2: 23.

Heaney, Seamus. 1990. *Selected Poems, 1966–1987*. New York: Noonday Press, Farrar, Straus and Giroux.

Heim, Michael. 1993. *The Metaphysics of Virtual Reality*. New York: Oxford University Press.

————. 1998. *Virtual Realism.* <http://www.mheim.com/rio>.

Irigaray, Luce. 1985. *This Sex Which Is Not One.* Trans. Catherine Porter with Carolyn Burke. Ithaca: Cornell University Press.

Jackson, Peter. 1989. *Maps of Meaning.* London: Routledge.

Jackson, Shelley. 1995. *Patchwork Girl, or, A Modern Monster.* Computer disk. Storyspace for Macintosh and Windows. Watertown, Mass.: Eastgate Systems.

Johnson, Steven. 1998. "The Alexa Effect." *Feed.* January. <http://www.feedmag.com/html/feedline / 98.06johnson/ 98.06 johnson_master.html>.

————, ed. 1995. "Page versus Pixel: The Cultural Consequences of Electronic Text." *Feed.* June. <http://www.feedmag.com /95.05dialog1.html>.

Joyce, Michael. 1995. *Of Two Minds: Hypertext Pedagogy and Poetics.* Ann Arbor: University of Michigan Press.

————. 1996a. "(Re)Placing the Author: "A Book in the Ruins." In *The Future of the Book,* ed. Geoffrey Nunberg and Patricia Violi. Berkeley and Los Angeles: University of California Press.

————. 1996b. *Twelve Blue. Postmodern Culture* 7, no. 3. <http://www.eastgate.com/twelveblue>.

Kac, Eduardo, ed. 1996. Special issue "New Media Poetry: Poetic Innovation and New Technologies." *Visible Languages* (Rhode Island School of Design) 30, no. 2.

Kauffman, Janet. 1993. *The Body in Four Parts.* St. Paul: Graywolf Press.

Kline, David. 1995. "Savvy Sassa." *Wired* 3, no. 3: 110–13.

Kristeva, Julia. 1991. *Strangers to Ourselves.* Trans. Leon S. Roudiez. New York: Columbia University Press.

Kunze, Donald. 1995. "The Thickness of the Past: The Metonymy of Possession." *Intersight3.* <http://www.arch.buffalo.edu/~intrsght /archives/intersight3/kunze/kunze.html>.

Kwinter, Sanford. 1992. "Landscapes of Change: Boccioni's *Stati d'animo* as a General Theory of Models." *Assemblage* 19:55–65.

Landow, George P. 1997. *Hypertext 2.0: The Convergence of Contemporary Critical Theory and Technology.* Baltimore: Johns Hopkins University Press.

Landow, George P., and Jon Lanestedt. 1991. *"In Memoriam" Web.* Storyspace hypertext software. Watertown, Mass.: Eastgate Systems.

Lanham, Richard. 1993. *The Electronic Word: Democracy, Technology, and the Arts.* Chicago: University of Chicago Press.

Leeser, Tom, and Alison Saar. 1997. *The Kunst Brothers.* <http://home.earthlink.net/~kunst/>.

Levy, David M., and Catherine C. Marshall. 1995. "Going Digital: A Look at Assumptions Underlying Digital Libraries." *Communications of the ACM* 38, no. 4: 77–84.

Lippard, Lucy. 1983. *Overlay.* New York: Pantheon.

Loader, Jane. 1997. *Flygirls.* <http://www.flygirls.com/>.

Lopez, Barry. 1990. "Losing Our Sense of Place." *Teacher,* February, 188.

Lunenfeld, Peter. 1998. "Demo or Die." <nettime>. Thu, 30 Jul 1998 12:21:26—0800, <http://www.desk.nl/~nettime/>.

Malin, Heather. 1999. "Contour and Consciousness." *Eastgate Quarterly Review of Hypertext,* forthcoming. Watertown, Mass.: Eastgate Systems.

Malloy, Judy. 1993. *Brown House Kitchen.* <http://www.artswire.org/~jmalloy/richmond/kitchen.html>.

Mangan, Katherine S. 1994. "A&M's Alchemy Caper." *Chronicle of Higher Education,* January 19, A19.

Marcel, Gabriel. 1963. "Value and Immortality." In *Homo Viator,* trans. Emma Craufurd. New York: Harper and Brothers.

Marker, Chris. 1992. *La Jette: Ciné-roman.* New York: Zone.

Maturana, Humberto. 1991. "Response to Jim Birch." *Journal of Family Therapy* 13:375–93.

McLaughlin, Tim. 1996. *25 Ways to Close a Photograph.* <http://www.knosso.com/nwhq/tim/tim_25.html>.

McLaughlin, Tim, Thomas Bessai, Maria Denegri, and Bruce Haden. 1997. *Light Assemblage.* <http://www.knosso.com/nwhq/blueprints/>.

Merleau-Ponty, M. 1962. *Phenomenology of Perception.* Trans. C. Smith. London: Routledge and Kegan Paul.

Merton, Thomas. 1956. *Thoughts in Solitude.* Boston: Shambala Pocket Classics.

———. 1983. *Woods, Shore, Desert: A Notebook, May 1968.* Santa Fe: Museum of New Mexico Press.

Milosz, Czeslaw. 1987. *Conversations with Czeslaw Milosz.* Ed. Ewa Czarnecka and Aleksander Fiut, trans. Richard Lourie. New York: Harcourt Brace Jovanovich.

———. 1988. *The Collected Poems.* Trans. Robert Hass, Robert Pinsky, with the author and Renata Gorczynski. New York: Ecco Press.

Mitchell, W. J. T. 1994. *Picture Theory: Essays on Verbal and Visual Representation.* Chicago: University of Chicago Press.

Mola Group [Carolyn Guyer, Michael Joyce, Nigel Kerr, Nancy Lin, and Suze Schweitzer]. 1995. *MOLA, World3.* Archived at <http://iberia.vassar.edu/mola/>.

Molinari, Guido. 1976. "Color in the Creative Arts." In *Ecrits sur l'art (1954–1975).* Ottawa: GNC.

Moulthrop, Stuart. 1991. "Reading from the Map: Metonymy and

Metaphor in the Fiction of *Forking Paths.*" In *Hypermedia and Literary Studies,* ed. Paul Delany and George Landow. Cambridge: MIT Press.

Mouré, Erin. 1988. *Furious.* Toronto: Anansi.

———. 1989. *WSW (West South West).* Montreal: Véhicule Press.

———. 1993. "'And Just Leave Them There, and Let Them Resonate': An Interview with Erin Mouré." Interview by Nathalie Cooke. *ARC31,* Autumn, Ottawa.

Murray, Janet. 1997. *Hamlet on the Holodeck.* Cambridge: MIT Press.

Nesbitt, Molly. 1992. "In the Absence of the Parisienne . . ." In *Sexuality and Space,* ed. Beatrice Colomina. Princeton: Princeton Architectural Press.

Novak, Marcos. 1996. *Trans Terra Form: Liquid Architectures and the Loss of Inscription.* <http://www.t0.or.at/~krcf/nlonline/nonMarcos.html> or <http://flux.carleton.ca/sites/projects/liquid/novak1.html>.

Nunberg, Geoffrey, ed. 1996. *The Future of the Book.* Berkeley and Los Angeles: University of California Press.

Olson, Charles. 1947. *Call Me Ishmael.* San Francisco: City Lights.

———. 1970. *The Special View of History.* Ed. Ann Charters. Berkeley, Calif.: Oyez.

———. 1974. *Additional Prose.* Bolinas, Calif.: Four Seasons.

Ottinger, Didier. 1996. "The Spiritual Exercises of René Magritte." In *Magritte.* Montreal: Museum of Fine Arts.

Page, Barbara. 1996. "Women Writers and the Restive Text: Feminism, Experimental Writing, and Hypertext." *Postmodern Culture* 6, no. 2. <http://jefferson.village.virginia.edu/pmc/issue.196/page.196.html>.

Paul, Sherman. 1976. *Repossessing and Renewing: Essays in the Green American Tradition.* Baton Rouge : Louisiana State University Press.

———. 1981. *The Lost America of Love: Rereading Robert Creeley, Edward Dorn, and Robert Duncan.* Baton Rouge: Louisiana State University Press.

———. 1992. *For Love of the World: Essays on Nature Writers.* Iowa City: University of Iowa Press.

Perrella, Stephen. 1996. B e i n g @ H o m e . . a s   B e c o m i n g   I n f o r m a t i o n   a n d   *H y p e r s u r f a c e.* <http://www.mediamatic.nl/doors/doors2/perrella/perrella-doors2-e.html>.

Petry, Martha. 1992. "Permeable Skins." Special issue "After the Book: Writing Literature, Writing Technology." *Perforations* 3 (spring–summer): n.p.

Pickles, John. 1995. "Representations in an Electronic Age." In *Ground Truth: The Social Implications of Geographic Information Systems,* ed. John Pickles. New York: Gifford Press.

Pound, Ezra. 1934. *ABC of Reading.* New York: New Directions.

———. 1954. *The Cantos of Ezra Pound.* New York: New Directions.

Rheingold, Howard. 1994. "PARC Is Back!" *Wired* 2, no. 2: 90–95.

Richmond, Wendy. 1994. "Muriel Cooper's Legacy." *Wired* 2, no. 10: 184.

Rosello, Mireille. 1994. "The Screener's Maps: Michel de Certeau's 'Wandersmänner' and Paul Auster's Hypertextual Detective." In *Hyper/Text/Theory*, ed. George Landow. Baltimore: Johns Hopkins University Press.

Rosenberg, Jim. 1993. "Intergrams." Computer disk for Macintosh and Windows *Eastgate Quarterly Review of Hypertext* 1, no. 1.

———. 1996a. "The Structure of Hypertext Activity." In *Hypertext '96: The Seventh ACM Conference on Hypertext*. New York: ACM Press.

———. 1996b. "The Interactive Diagram Sentence: Hypertext as a Medium of Thought." *Visible Languages* (Rhode Island School of Design) 30, no. 2, 102–18.

———. 1997. *The Barrier Frames*. Computer disk. Watertown, Mass.: Eastgate Systems.

Rothenberg, Jerome, ed. 1968. *Technicians of the Sacred*. New York: Anchor.

Ryan, Marie-Laure. 1994. "Immersion vs. Interactivity: Virtual Reality and Literary Theory." *Postmodern Culture* 5, no. 1. <http://jefferson.village.virginia.edu/pmc/issue.994/ryan.994.html>.

Sahagún, Fray Bernardino de. 1961. "The People." In *Florentine Codex: General History of the Things of New Spain*. Trans. and ed. Arthur J. O. Anderson and Charles E. Dibble. Monographs of the School of American Research No. 14. Salt Lake City: University of Utah Press.

Sainsbury, Alison. 1994. Email, October 11.

———. 1995a. "Maps and Gaps: Storyspace and the Geography of Gender." Paper presented to Conference on College Composition and Communication, Washington D.C., March.

———. 1995b. Notes for lecture, "Hypertextual Ladders," Vassar College.

Sanford, Christy Sheffield. 1997a. *The Landscape/Scene/Web*. <http://gnv.fdt.net/~christys/land-web/landscape.html>.

———. 1997b. *Solstice*. <http://gnv.fdt.net/~christys/light/moving/htm>.

Sauer, Carl O. 1968. *Northern Mists*. San Francisco: Turtle Island Foundation.

Schweitzer, Suze. 1994. "View from Prague." Email, January 11.

Serres, Michel. 1982. "The Origin of Language: Biology, Information Theory, and Thermodynamics." In *Hermes; Literature, Science, Philosophy*, ed. Josué V. Harari and David F. Bell. Baltimore: Johns Hopkins University Press.

Shklovsky, Viktor. 1990. *Theory of Prose*. Trans. Benjamin Sher. Elmwood Park, Ill.: Dalkey Archive Press.

Snyder, Ilana, ed. 1997. *Page to Screen: Taking Literacy into the Electronic Era*. Melbourne: Allen and Unwin.

Soja, Edward. 1996. *Thirdspace.* Cambridge, MA: Blackwell.

———. 1993. "History: Geography, Modernity." In *The Cultural Studies Reader,* ed. Simon During. New York: Routledge.

Sontag, Susan. 1966. *Against Interpretation, and Other Essays.* New York: Farrar, Straus and Giroux.

Stein, Gertrude. 1933. *The Autobiography of Alice B. Toklas.* New York: Harcourt, Brace and World.

———. 1934. *The Making of Americans.* New York: Harcourt, Brace and World.

———. 1990. "Composition as Explanation." In *Selected Writings,* ed. Carl Van Vechten. New York: Vintage.

Stivale, Charles J. 1995. "Works and Days Update." Email, March 27.

Syverson, Peg, Carolyn Guyer, Marjorie Luesebrink, and Michael Joyce. 1997. "Walk Four Ways One Time: Narrative Coherencies." *Pre/Text.* 16:1–2, 70–96.

Taylor, Mark C., and Jack Miles. 1998. *Hiding.* Chicago: University of Chicago Press.

Taylor, Mark C., and Esa Saarinen. 1994. *Imagologies: Media Philosophy.* New York: Routledge.

Thoreau, Henry David. 1997. *Walden.* Public domain TEI edition prepared by the North American Reading Program. <http://dev.library.utoronto.ca/utel/ nonfiction/thoreauh_wald /wald_all.html>.

Tikka, Heidi. 1994. "Vision and Dominance—a Critical Look into Interactive System." In *ISEA '94 Proceedings: The Fifth International Symposium on Electronic Art,* Inter-Society for the Electronic Arts, Helsinki, August 20–25. <http://www.uiah.fi/bookshop /isea_proc/nextgen/j/10.html>.

Ulmer, Gregory L. 1989. *Teletheory: Grammatology in the Age of Video.* New York: Routledge.

Wagner, Annette, and Maria Capucciati. 1996. "Demo or Die: User Interface as Marketing Theatre." In *CHI 96—Electronic Proceedings,* ed. Ralf Bilger, Steve Guest, and Michael J. Tauber. <www.acm.org/sigs/sigchi/chi96/proceedings /desbrief /Wagner /aw_txt.htm>.

Wallen, Jeffrey. 1997. Introduction to *Who's Who in Central and East Europe, 1933.* <http://www.uni-lueneburg.de/memory/>.

Wardrip-Fruin, Noah, with Michael Crumpton, Nathan Fruin, Kristin Holcomb, Kirstin Kantner, Chris Spain, and Duane Whitehurst. 1997. *Gray Matters.* <http://www.cat.nyu.edu/graymatters/>.

Weinbren, Grahame. 1995. "In the Ocean of Streams of Story." <http://www.sva.edu/mfj/gwocean.html>. Also published in *Millenium Film Journal* 28 (spring): Interactivities, 15–30.

Wenders, Wim. 1991. *The Logic of Images: Essays and Conversations.* London: Faber and Faber.

Wenders, Wim, and Peter Handke. 1987. *Wings of Desire* (Der Himmel über Berlin). Berlin: Road Films.

———. 1988. *Der Himmel Über Berlin: Ein Filmbuch.* Frankfurt am Main: Suhrkamp.

Whitman, Walt. 1965. *Leaves of Grass.* Ed. Harold W. Blodgett and Sculley Bradley. New York: New York University Press.

Wortzel, Adrianne. 1997. *Ah, Need.* <http://www.artnetweb.com /projects/ahneed/ahneed.html>.